UNDERSTANDING
MAGNETISM
MAGNETS, ELECTROMAGNETS
AND
SUPERCONDUCTING MAGNETS

No. 2772
$17.95

UNDERSTANDING

MAGNETISM

MAGNETS, ELECTROMAGNETS
AND
SUPERCONDUCTING MAGNETS

ROBERT WOOD

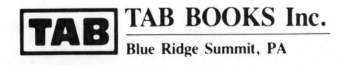

TAB BOOKS Inc.

Blue Ridge Summit, PA

Notice

It is the intention of the author that the information presented is accurate and stresses the need for safety, however; due to the nature of electricity, neither the author nor TAB BOOKS Inc. is liable with respect to the use of the information herein.

FIRST EDITION
FIRST PRINTING

Copyright © 1988 by TAB BOOKS Inc.
Printed in the United States of America

Library of Congress Cataloging in Publication Data

Wood, Robert W., 1933-
Understanding magnetism : magnets, electromagnets, and
superconducting magnets / by Robert W. Wood.
p.   cm.
Includes index.
ISBN 0-8306-0772-2   ISBN 0-8306-2772-3 (pbk.)
1. Magnetism—Popular works. I. Title.
QC753.5.W66   1988
538—dc19   88-8559
                                                          CIP

Questions regarding the content of this book
should be addressed to:

Reader Inquiry Branch
TAB BOOKS Inc.
Blue Ridge Summit, PA 17294-0214

Contents

Acknowledgments

It is with sincere appreciation that I thank the following people and businesses for their contributions in the writing of this book: Arizona Public Service Company, Phoenix, Arizona; British Broadcasting Corporation, London, England; Jess Castellano, Mesa, Arizona; Delco Remy, Anderson, Indiana; Motorola Semiconductor Group, Phoenix, Arizona; Salt River Project, Phoenix, Arizona; Sam W. Taliaferro VII, Magnetic Engineering, Atlanta, Georgia; Varian Semiconductor Equipment Group, Tempe, Arizona; Westinghouse Electric Corporation, Boise, Idaho and Orlando, Florida.

Introduction

Magnetism is a fascinating phenomenon. The invisible force surrounds us throughout our lives, yet we seldom appreciate its wide influence.

This book was written to show some of the uses man has found for magnetism, and the close relationship magnetism has with electricity. Most of the material available on either subject avoids this strong connection.

Discoveries made in the last two decades indicate that even Earth's magnetic field might have more influence on life than we ever dreamed.

Although technical passages are included, this book is written for non-technical people who want a deeper understanding of how magnetic fields affect everything from the northern lights to homing pigeons.

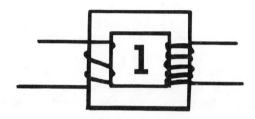

What is Magnetism?

Magnetism is an intriguing phenomenon that has been puzzling mankind since the first magnetic stones were discovered in an ancient area in Asia Minor called Magnesia. The curious relationship between magnetism and electricity is the keystone for the building blocks to this fascinating world around us.

Youngsters and adults alike can become engrossed for hours with a couple of magnets. It soon becomes obvious there is some kind of force field around these magnets. You can easily see what they do and their attraction or repulsion for each other, but it is far from obvious how they do it. Magnetism has force and power. It can do work. People use it every day, while scientists are still trying to unravel its mysteries. For example many people have a little flower or a butterfly stuck on their refrigerator with a magnet to hold notes. The next time you drive by a wrecking yard, take a good look at the crane that moves large pieces of metal from one place to another. When your telephone rings, that also demonstrates an effect of magnetism.

The next time you're watching television, try to imagine the scene you're looking at is being created by a single dot racing back and forth across the front of your picture tube. This dot is directed and controlled by a magnetic field created by a deflection yoke on the back of your picture tube.

Magnetism is the physical phenomenon of a magnetic field; it can be loosely defined as an invisible force of attraction. Most people tend to think of magnetism as something in and around an iron bar that allows it to pick up paper clips or nails, but with just a little research, you'll soon discover that you spend your entire life enveloped in this mysterious, invisible force.

THE NORTHERN LIGHTS

Magnetic storms and the aurora borealis, normally thought of as the northern lights, are directly related to this phenomena. To better understand this, you need to take a look at Earth and the area known as the magnetosphere from a distant position in outer space (Fig. 1-1). From this imaginary position in space you can see that the magnetosphere begins at an altitude of about 62 miles above Earth and extends outward to a region called the magnetopause. The shape of the magnetosphere is somewhat blunt on the side facing the sun but has a long tail, much like that of a comet, extending from the opposite, or shaded, side of Earth. The blunt shape is caused by the pressure felt by the magnetopause from the solar wind; a hot particle-laden gas speeding from the sun at almost one million miles per hour.

On the upwind side, facing the sun, the magnetosphere extends outward about 10 R_e. What is R_e? (R_e = Earth radius = about 3,960 miles) Simple mathematics indicate that our magnetosphere extends out about 39,600 miles

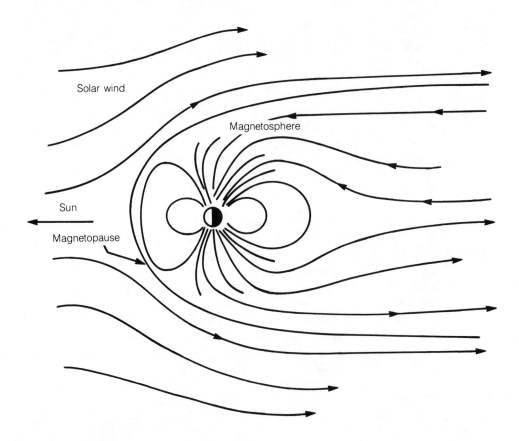

Fig. 1-1. The magnetosphere shielded from the solar wind by the magnetopause.

from Earth. Because Earth tends to expand or flex its magnetic muscles from time to time, this distance can vary between 6 and 14 R_e. Over the poles, the distance is about 15 R_e. On the downwind side, the shape of the tail is not determined by pressure from the solar wind (if it were it could expand into the shaded or protected areas); its shape is determined solely by the strength in the magnetic field of the tail itself.

This powerful field contains all the magnetic field lines that enter or leave from an area within about 930 miles of Earth's magnetic poles. In the southern (bottom) half of the tail, the direction of the lines of the magnetic field are away from Earth. In the northern half they complete the loop or circuit and are going toward Earth. These two halves are isolated from each other by a thin neutral barrier of a low magnetic strength containing a plasma. This plasma has enough pressure to prevent the linkage of the top and bottom halves, like the traffic barriers on a freeway.

In 1981, the Dynamics Explorer 1 was launched into a 15,500-mile-high polar orbit from Vandenburg Air Force Base in California. This satellite took some 200,000 photographs of the polar auroras. Information presented during a February 1985 international conference on solar wind interaction with Earth's magnetic field at the Jet Propulsion Laboratory in Pasadena, California indicated that researchers had recently discovered a huge energy zone in the tail.

Lou Frank, a University of Iowa physicist, announced that researchers analyzing some of the photographs had located this large energy zone some 400,000 miles from our planet. This energy zone, it turns out, is the power supply for the northern lights (aurora borealis) and the southern lights (aurora australis).

This invisible, egg-shaped power source, made up of electrically charged particles, is 20 to 30 times larger than Earth and located about 400,000 miles downstream on the side away from the sun. Particles from the solar wind are trapped in this energy zone, and then through complex reactions, the voltage in this power source accelerates these particles into Earth's polar regions.

As these electron and proton particles approach Earth, they are traveling at about 20,000 miles a second. At that speed, it would take about a second and a half to go around the equator. These particles are packing energies of more than 10,000 electron volts. This is more than enough energy to ionize atoms.

Most of the time they travel high above Earth's surface in an almost complete vacuum with little chance of a collision with an air molecule. However, as the protons and electrons approach Earth's magnetic poles at a lower altitude where the air is more dense, collisions occur (Fig. 1-2). In these collisions, atoms of atmospheric oxygen and nitrogen are ionized. Ionization takes place when a violent impact changes the number of electrons normally found in the atom. When these ions recombine, releasing the energy they absorbed during the collision, light is emitted. This effect creates our mystic northern lights. This unearthly display of colors from yellow-green to reddish purple are usually only visible

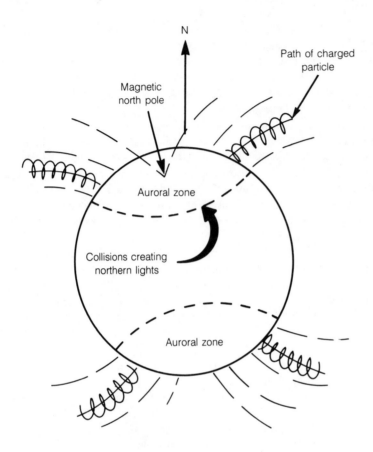

Fig. 1-2. The auroral zone is the region where electrons colliding with air molecules produce visible light displays.

from the higher or northern latitudes. Visual displays can range from a red glow to a series of green curtains 60 to 70 miles tall, waving and drifting across the sky.

Magnetic storms are really not a weather phenomena but occur when sun spots and giant flares erupt on the sun. As a result the solar wind becomes much stronger. For a few days the aurora is much brighter and the entire upper atmosphere becomes very disturbed. Radio and telephone communications are plagued with static. Television reception becomes distorted and even a magnetic compass can behave strangely.

EARTH'S MAGNETIC FIELD

It was William Gilbert (1540-1603), a physician to England's Queen Elizabeth I, who first began to use scientific methods to study magnetic phenomena. In 1600, Gilbert published his discovery that Earth itself is a magnet. We do indeed

4

live on a huge magnet, but Earth's magnetic field must be created, or generated, by something.

For a closer inspection of Earth, you can bore down to the very center, and discover a rotating, solid sphere about 1,600 miles in diameter (Fig. 1-3). From the habits of volcanoes, scientists know that the interior of the Earth is hot. Temperature measurements taken in mines show an increase in temperature of about 1 degree centigrade for each 100 feet of depth.

Pressure also increases with depth; it is estimated in excess of 3 million atmospheres at the center of Earth. In physics, atmosphere is a unit of pressure equal to 14.69 pounds per square inch.

Due to the extreme temperatures and pressure, this inner core is believed to be a dense liquid metal made up of nickel and iron. This sphere can rotate because it is suspended in a fluid substance about 1,400 miles thick. This outer core is made up of a mixture of basic rock minerals and metallic iron, with the metallic iron increasing with depth. The friction generated between the rotating inner core and outer core generates heat. This keeps the outer core fluid but also generates a great deal of pressure on the crust.

The crust is about 30 miles thick and is essentially solid, beginning mainly with sedimentary rocks for the surface changing to a zone of granite and gneiss (rock similar to granite). The mantel begins below this region. It is an 1,800-mile-

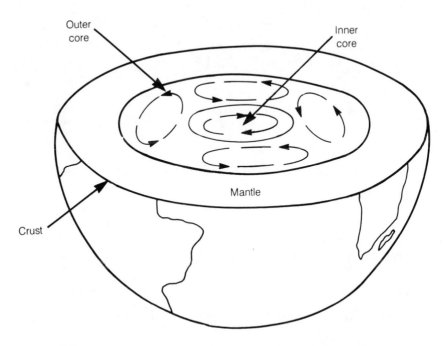

Fig. 1-3. The three spheres of Earth: The inner core, about 1,600 miles in diameter; the outer core, 1,400 miles thick; and the mantle, 1,800 miles thick.

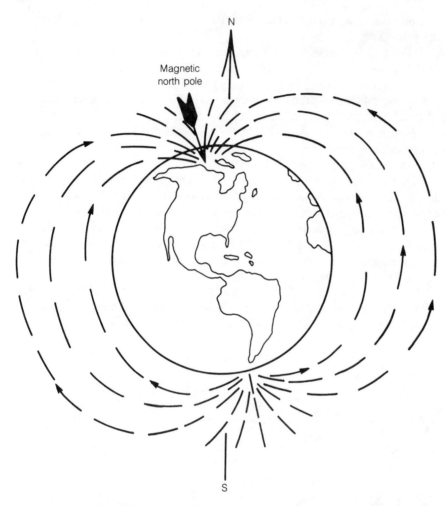

Fig. 1-4. The Earth's magnetic axis is not the same as the true north-south axis. The magnetic north pole is located at 76 degrees north latitude; 102 degrees west longitude. The true north pole is 0 degrees latitude by 0 degrees longitude.

thick zone of basic rocks containing a great deal of iron in silicate minerals but no iron in a metallic state. As this pressure builds, it is bled off through pressure relief valves in the crust as volcanic eruptions.

The effect created by this inner sphere rotation is very similar to the principles of a generator in that it creates electric current. The study of magnetohydrodynamic motions indicates that we are living on a huge dynamo. Somewhat unnerving, isn't it?

One of the first things you learned about this push-pull force of magnets is that like poles repel and unlike poles attract. For example the north-pointing

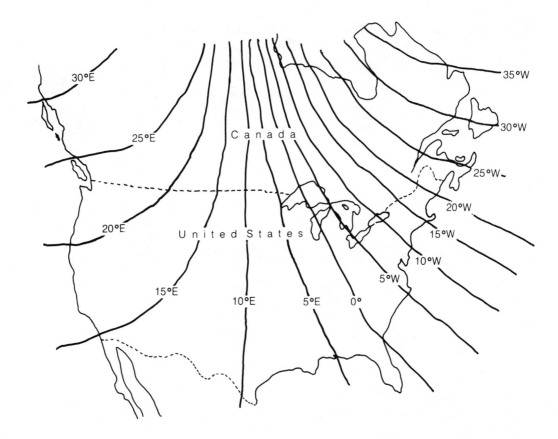

Fig. 1-5. Magnetic declination lines across Canada and the United States.

pole of a magnetic compass is actually the south pole of the needle. It is labeled north because it points north, but it's actually the south pole because the needle itself is simply a long, thin magnet, balanced on a pivot point that allows it to easily rotate and align itself with the north-south lines of Earth's magnetic field.

Earth's magnetic north and south poles are not located at the same place as the true, or geographic, north and south poles (Fig. 1-4). The north pole that attracts a magnetic compass is located near Prince of Wales Island in Canada, about a thousand miles from the true north pole. It is not even on the surface, but is buried about 70 miles inside the Earth. The magnetic south pole is about 1,500 miles from the true south pole directly south of Sydney, Australia.

Magnetic compasses rarely point true north because of these differences. Early navigators soon learned that they had to know how far from true north their magnetic compasses were pointing. This angle between the two north poles is called the angle of magnetic declination (Fig. 1-5). This angle, measured in degrees East or degrees West, is added or subtracted from the compass heading

7

to give a true heading. (Degrees East are subtracted, degrees West are added.) The magnetic north pole tends to wander somewhat, causing these angles to change, so charts showing magnetic declination angles must be upgraded every few years.

Birds take advantage of Earth's magnetic field in their seasonal migrations. A very small magnetic crystal located between the brain and the skull of pigeons, discovered by Charles Walcott, Jim Gould, and Joe Kirschvink, strongly suggests that birds use the geomagnetic field for orientation.

F. W. Merkel and W. Wiltschko, working in Frankfurt, Germany with a team of enthusiasts, proved to ornithologists that birds are definitely affected by magnetic fields. In 1978 they introduced the ''V.W. effect.'' Birds carried to a release site in the back of their Volkswagen pickup sometimes were confused and could not orient themselves when they were released. Other experimenters were having similar problems. It was soon proven that the close proximity of the cages to the engine's generator was interfering with the birds' magnetic reception on the way to the release site.

It has also been proven that solar flares, which cause fluctuation in Earth's magnetic field, tend to disrupt birds' abilities to navigate. They do seem to take compass bearings or magnetic headings, but not in the same way pilots or mariners use a compass. Their ability to detect magnetic directions is not quick; direction finding can take 15 minutes or more. Also, birds seem to sense the angle the magnetic lines of force strike Earth instead of the horizontal way we use a compass (Fig. 1-6).

At the northern and southern magnetic poles the magnetic lines of force are vertical. Halfway between them near the equator the force is horizontal. The field drifts westward each year, probably because the core is rotating eastward a little slower than the outer more solid part of Earth. It is thought that alterations in the circulation of the molten material in Earth's core are responsible for some of the changes in Earth's magnetic field and could even slightly affect the rotation of our planet. It follows that this would in turn affect the circulation of the oceans along with our atmosphere.

There also exists a pattern of regular cycles in Earth's magnetic field, predictable twenty-four hour periods where about mid-day every day the field strength rises and then falls.

Europe has an extensive network of magnetic observatories where sensitive instruments are used to keep a constant surveillance on Earth's magnetic field. Dr. Stuart Malin, working at one such observatory, has found that this pattern of variations changes slightly from summer to winter with the winter having smaller peaks in the field. When this information is fed into a computer, other cycles begin to appear, including monthly cycles that are in phase with the ocean's tides. These cycles are not caused by Earth's core but come from a weaker source. They seem to be the effect of a system of electrical currents flowing in the ionosphere. As this current varies from night to day along with the seasons of the year, it tends to be supported by an even weaker source operating within

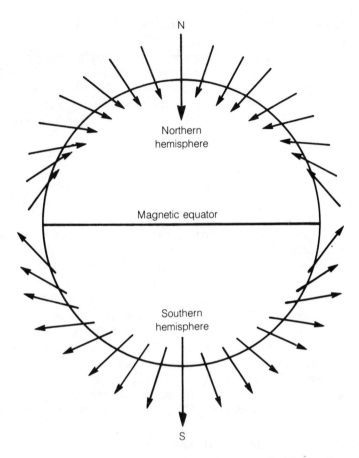

Fig. 1-6. Magnetic lines entering or leaving Earth at an angle. This angle varies with latitude.

the oceans. Dr. Malin's research suggests that Earth's magnetic field is a complex product resulting from all of the electrical activity within and around our planet. If this is true, there should be some connection between this force field and our weather.

For some time investigators have wondered if there was some connection between solar flare-ups and the activity in our lower atmosphere. One of the first scientists to look for this association was the British meteorologist Dr. Stagg. He found a distinct difference in the average pressure variations between the magnetically-disturbed days, during magnetic storms, and normal magnetically-quiet days.

Dr. Joe King from the Appleton Laboratory in Slough (southeast England) has completed research agreeing with Dr. Stagg. There is evidence that fluctuations in Earth's magnetic field are closely associated with our weather patterns. The fluctuations don't cause a change in the weather; however, changes in the magnetic field correspond to weather conditions because both are the consequence of enhanced solar activity.

In some circles of meteorological science, the notion that these insignificant variations could influence the weather is viewed with disbelief. Dr. Goesta Wallin at the Lamont-Doherty Geological Observatory in New York, however, has further evidence that it does. Dr. Wallin has devoted years of research to the connection between magnetism and weather and he concludes the evidence is clear for anyone to see. Curves representing the lag between the magnetic variations and the average temperature and rainfall that occurred several years later suggest astonishing capabilities for long-range forecasting.

Because weather is the result of many complicated variables, Dr. Wallin does not imply that magnetism is the only influence, but claims that it is the principle influence of climate.

How could long-term fluctuations in Earth's magnetic field cause parallel variations in our weather?

Dr. King feels that charged particles coming from the sun affect the nitric oxide in our atmosphere. This influences the quantity of ozone which, in turn, has an affect on the temperature in the stratosphere. At the same time, charged cosmic rays coming toward Earth affect the electricity in the atmosphere. When these charged particles enter Earth's magnetic field, the field strongly affects their distribution over the surface of Earth. This could be a major factor in the weather patterns that are experienced around the world.

Solar eruptions also affect or disturb Earth's magnetism. A severe weather phenomenon active in the equatorial Pacific, known as El Niño, occurred following such sudden changes in 1956, 1962, 1967, 1970, 1974, and 1981.

MAGNETS AFFECT PEOPLE

Magnetic forces affect humans in many ways. For example, some people appear to have an unconscious ability to detect magnetic bearings. Dr. R. Robin Baker, a faculty member at the University of Manchester in England, has done extensive research in the field of human navigation. On June 29, 1979 low heavy clouds delivered a steady drizzle on a group of students as they boarded a bus in the parking lot at Barnard Castle. Soon after the students were seated, they blindfolded themselves and each one put on a cap with a magnet attached to it. They were instructed to just sit back, relax, and concentrate.

The driver engaged the clutch and began to work his passengers through the town, emerging on a reasonably straight road to the southwest. After a period of time, when it had gone about 6 miles or so, the bus stopped. The students were asked to write down on a card the compass bearing back to the school. The bus started up again, turned eastward and drove a few miles to another stop. The sequence was repeated. Then the cards, blindfolds and magnets were collected.

Unknown to them, only half of the students wore magnets; the other half were using non-magnetic brass bars of the same size and weight. The results of the experiment were dramatic. The ones wearing the brass bars had written close approximate magnet headings of the direction back to the school while the students wearing magnets could not. The students not wearing the magnets seemed to actually be using Earth's magnetic field for direction finding, as they certainly did not have any help from the sun on such a day.

It is generally thought that the magnetic properties in matter are caused by the orbital motions and spins of the electrons in the atoms of that particular matter. The spinning electron in its orbit creates a tiny electric current loop, or an electric circuit with current flow, and whenever there is current flow, a magnetic field exists.

Molecules in an iron bar are normally arranged in a jumble. In this position their attraction for each other is balanced. One cancels out another, but once the molecules are aligned in a north-south direction, something strange happens: they pull as a unit. However, when you have opposite poles at each end, somewhere near the middle things have to change. The molecules in the center cancel each other out so the middle of a bar magnet has no magnetization, but when a bar magnet is cut in half it still retains its poles. This fact offers support to the molecular theory of magnetism and the orbital spin of electrons.

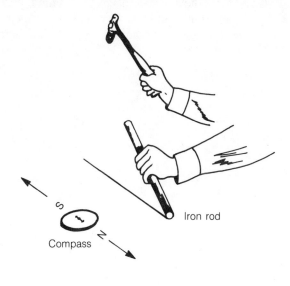

Iron rod

Compass

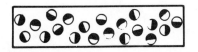 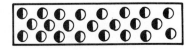

Fig. 1-7. Aligning the molecules in an iron rod north and south to make a magnet.

This theory can be further demonstrated by making a magnet with a sharp rap on an iron rod and aligning the molecules north and south. This homemade magnet can be made with a compass, an iron rod and a hammer (Fig. 1-7). Using the compass, locate a north-south line in the dirt. Hold the iron rod in one hand over the line with the north end angled down and in the ground slightly. With a sharp rap, hit the south end with the hammer. All you want is a sudden jar to align the molecules inside the iron bar. If you were successful you now have a magnet.

Molecules are set in motion by heat. If you heat a magnet, the molecules move out of their north-south line and you no longer have a magnet. A sudden jar can also demagnetize a magnet by knocking the molecules out of line.

Earth's magnetic field should not be confused with Earth's gravity. Where gravity is only concerned with the mass or weight of an object and affects everything, magnetism has an intimate relationship with the electron and electricity and is highly selective. There is some relationship, however, as our mind operates primarily through electrical circuitry but can also be effected by phases of the moon and gravity. Mental patients tend to become more disturbed during a full moon, and gravitational pull causes the ebb and flow of the tides in the oceans.

This exciting, invisible force field called magnetism offers unlimited possibilities because of its relationship with electricity. Future developments in science will have to deal with one or the other—more probably, both.

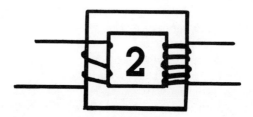

History of Magnetism

The ability to generate large amounts of electricity has provided us with a very comfortable way of life. Satellite communications, television, heating and cooling systems in our homes, even medical diagnosis are made easier with magnetism and electricity. The electronics that make possible computers, inexpensive miniature calculators available in a wrist watch, and the space shuttle is almost taken for granted. These electrical and electronic marvels of today all began with the curiosity, and consequently the study, of magnetism some 2,500 years ago.

EARLY HISTORY

Legend says that the word magnetism comes from the name of a shepherd boy in ancient Greece. Magnus, the shepherd, was tending his flock on Mt. Ida when he accidentally placed the tip of his staff on a large stone. The stone exerted so much pull that Magnus was unable to free his staff.

More probably, however, the name was derived from the strange stones discovered in an ancient country of Asia Minor called Magnesia. These magnet stones were actually a type of iron ore that are now called magnetite.

Many strange beliefs surrounded these magnetic stones. They were thought by some to be able to cure rheumatism, cramps, and the gout, and when placed on top of the head they allowed the wearer to hear the voices of the gods. A paste made of powdered magnetite and oil was a sure cure for hair loss, and when worn on a charm or ring would attract a lover.

In the seventh century B.C., a Greek philosopher and mathematician named Thales was the first to record his observations of the curious phenomena of magnetism and static electricity. He found that if he rubbed amber with a piece of wool it would pick up light objects, feathers, straw and bits of dried grass. Thales concluded that rubbing the amber made it magnetic. He noticed that the rubbed amber, however, would not attract metal. His other experiments showed that he could pick up small bits of iron with a piece of magnetite without rubbing it.

We now know that rubbing the amber created a static electrical charge, and not magnetism. A similar static charge can be built in your body by walking across a nylon or wool rug. This energy is discharged with a brief spark when you touch a doorknob or another person. Hundreds of years would pass before man would learn to use this great tool.

The loadstone was discovered when someone noticed that one end of a piece of magnetite was attracted by Earth's north pole. When this stone was suspended by a string it would turn until it was pointing north and south.

About 376 B.C. a Chinese general named Haung Ti was the first to employ a loadstone as a compass. Chinese generals and military commanders during the Han dynasty were the first to use an actual magnetic compass; however, for some reason it was used only on land. Couldn't a ship fitted with such a device be navigated even when the sun and the stars were hidden by clouds?

Not until 900 years later, in the thirteenth century, did Chinese navigators take their prize to sea. Arab sailors soon adopted this wonder and brought it to Europe. The magnetic compass was the key that opened the door to the world for European exploration. Certainly Christopher Columbus made good use of the magnetic compass, and he noted in his records that the magnetic north differed from true north. This magnetic deviation, which was already known to navigators, was called *declination*. Not much progress was made in understanding magnetism until William Gilbert came along.

ELECTRICITY AND MAGNETISM: STUDIES BEGIN

Gilbert spent 12 years at St. John's College, Cambridge and graduated in 1560. He kept up his technical studies, and became mathematics examiner at the Royal College of Surgeons in 1565. Gilbert received his doctorate in 1569 and by the mid-1570's had established a prosperous practice in London. As he rose in his profession he continued his private studies and experiments while attending to the aristocracy and mingling with the intellectuals. These studies produced his greatest contribution to the science of magnetism: his book, *De Magnete*, which was published in 1600.

England had just defeated Spain's "Invincible Armada" and Her Majesty, Queen Elizabeth I was acquiring a strong interest in science. She had the most powerful navy in the world and realized that navigation would play a critical part in commerce and in acquiring colonies as well as for ocean supremacy. (Coincidentally perhaps, 1600 was the same year that William Gilbert was appointed Royal Physician to the Queen.) England's great navigators Sir Francis Drake and Sir Walter Raleigh were also very interested in Gilbert's work with magnetism.

Although the ancients had known about amber's peculiar ability to pick up feathers and other light objects, this effect remained a curiosity until Gilbert began his studies and classified it as a science called electricity.

The word electricity comes from *elektron*, the Greek word for amber. Gilbert used the word *electrum*, the Latin word for amber, and made up the word *electrica* to cover the other substances that acted like amber. He separated electricity from magnetism.

His book *De Magnete*, the earliest work on magnetism, treated Earth, for the first time, as a huge magnet. This viewpoint provided a rational basis for the understanding of the movement of a compass needle and its attraction to Earth's north and south magnetic poles, a critical point in the advancement of navigation.

William Gilbert applied scientific methods to study this strange phenomenon and in his honor a unit of magnetic strength is now called a gilbert. However, man's knowledge of magnetism would advance very little for two more centuries.

Progress in the understanding of electricity, however, fared somewhat better. Otto Von Guericke (1602-1686) built and demonstrated the first electric generator in 1660 (Fig. 2-1). Guericke's machine consisted of a large glass ball containing sulfur mounted on a long shaft with a hand crank. The glass ball was rotated at high speed, a cloth was applied, and sparks would leap between a spark gap connected to two brushes that touched the spinning sphere.

Gilbert's studies had convinced him that Earth was a huge magnet rotating in space. However, many intellectuals of the time disagreed with him. They believed that Earth was the center of the universe; Earth was fixed and the sun and all the other planets revolved about it.

In 1629 Niccolo Cabeo published the book *The Magnetic Philosophy*, trying to prove Gilbert was wrong. He believed that when "an electric" (something that would hold a charge) was rubbed until it became hot, the heat would create a force that would push all the air away from it. This low pressure area would allow two objects to come together. His theory was debatable until 1675 when an Irish physician Robert Boyle (1627-1691) developed a vacuum pump efficient enough to prove that magnetism works just as well in a vacuum as it does at atmosphere.

Many experimenters of that time were working with something called a barometric light. A tube was partially filled with mercury, and then air was evacuated from the tube. The experimenters found that vigorously shaking the tube produced a glow or an occasional flash of light.

Francis Hauksbee (?-1713), experimentalist for the Royal Society in London, published a book in 1709 containing a variety of experiments. Hauksbee was very interested in Boyle's vacuum pump and the effect that air, or the lack of it, had upon magnetism. His endless experiments with equipment he invented

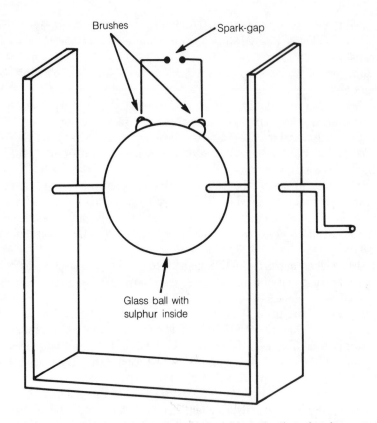

Fig. 2-1. Otto Von Guericke's generating machine, the first electric generator.

and built produced many electrostatic effects in evacuated spheres. Once he observed a weird glow while placing his hands on a rotating glass ball. He soon discovered that by varying the amount of vacuum, he could increase the level of light until it reached something very near an effect called St. Elmo's Fire. (St. Elmo's Fire is the name given to a strange glow that is sometimes seen around ships in a thunderstorm. It is actually an electrical charge created by the storm.)

In one experiment Hauksbee created a vacuum in a glass ball nine inches in diameter and set it to spinning. While rubbing it with a cloth (probably wool) a remarkable thing happened. He generated enough light in a darkened room to read by.

The glow in the barometric light was created because the tube built up an electrical charge from the friction of the mercury rubbing the glass as it was shaken, similar to neon lights today. Hauksbee suspected that the production of light and electricity from friction were somehow related. This effect he called *phosphorescence*. His discoveries were significant, for they challenged the beliefs of his time about electricity.

A simple way to duplicate Hauksbee's experiment is to rub a hollow tube enclosing a fluorescent one. While I didn't have these things handy, I did have a circular fluorescent tube in a desk lamp and a plastic salad bowl in the kitchen. Placing the tube in the bowl, turning out the lights, and rubbing the bottom of the bowl with the tail of an old Pendleton shirt produced a slight glow. When I touched the plug on the tube with my finger, a brief bright light would glow in the tube.

Another Englishman discovered that some substances conduct electricity and others do not. Stephen Gray was born between 1666 and 1696 and died in 1736. Gray wanted to see if electricity could be transmitted from one point to another. At first thought, this may sound very simple, but keep in mind the battery hadn't been invented yet and there was no sustained current to work with. They were left to deal with very small fluctuating currents of static electricity. Of course, some kind of volt meter would have been a big help too.

Gray placed an ivory ball on a 765-foot long piece of twine connected to a glass tube. His transmission line was suspended by silk threads. When the tube was rubbed, the ball at the far end would attract and then repel a feather, and would occasionally pick up thin pieces of brass. He then discovered that by wetting the twine he got even better results. Gray theorized that electrification was a fluid. He began substituting metal wires for the silk support threads, and the charge quickly dissipated. He realized that conductors would not hold a charge but non-conductors such as silk and glass would hold a charge. We now call these non-conductors *insulators*.

Then a Frenchman appeared, an ideal successor to Hauksbee and Gray. Charles F. Dufay (1698-1739) was already a leading member of the Paris Academy of Science at age 35. Dufay was a dedicated experimenter and interested in almost every science then known. He was also a member of the Royal Society in London. In late 1733, Dufay took on an assistant. He and J. A. Nollet, duplicating Gray's experiment, extended the range of conduction to 1256 feet, nearly a quarter of a mile.

In one experiment Dufay found that when he placed a thin sheet of gold leaf against his charged glass tube the sheet jumped from the tube and floated in the air. He then repeated the experiment except this time he introduced a ball of rubbed copal, a resin similar to amber. He slowly moved the copal near the floating gold leaf. They each had an electrical charge, so he reasoned that they should be attracted to each other. To his surprise, the gold sheet flew to the glass tube and stuck there.

He puzzled over the problem and concluded that he was dealing with two different electricities. Further experimenting indicated that first objects were attracted to the electrified glass tube, then some sort of communication took place, and then the objects were repelled. He had discovered that unlike objects attract and like charges repel. Dufay further reasoned that all bodies, everything, must contain some amount of electricity.

Human nature being what it is, this newfound science suddenly took on a carnival air. Some experimenters of the time began to exploit this new mystery. Frictional generating machines became larger. There were many public demonstrations, and much like the medicine shows of the old west, some of the "professors" assured an uneducated audience that this new miracle could cure all their medical ills. That seemed to be its only useful purpose at that time.

In general, these early pioneers thought that since electricity was a fluid, and fluids tend to evaporate, electricity would lose its charge, or dissipate, in the open air.

Ewald Jurgen von Kliest, the son of a Prussian official, wondered if you could put electricity in some kind of chamber or vessel and seal it from air so it would retain its charge a little longer. His experiments began with a glass bottle half-full of water and sealed with a cork. A nail was driven through the cork until it touched the water. He then picked up the bottle and brought the head of the nail close enough to his frictional machine to receive a charge. To see if he'd truly captured electricity in a bottle, he then brought the head of the nail near a non-electrified object. A healthy spark bridged the gap. His free hand then moved to touch the nail. His arm was suddenly jerked to his chest. He had, indeed, discovered that you could store electricity. He had invented the capacitor.

A few years later, in Leyden, Holland, Pieter Van Musschenbrock (1692-1761), an instructor of mathematics, involved himself with a very similar experiment and after receiving some very severe shocks, announced the invention of the Leyden Jar (Fig. 2-2).

BENJAMIN FRANKLIN

Benjamin Franklin (1706-1790), statesman and politician, is well remembered for flying a kite in a thunderstorm, but few people know that he invented the lightning rod, that sharp pointed spike that adorns rooftops throughout the south and midwest. Aside from being a printer and the author of *Poor Richard's Almanac*, he should also be considered America's first electrician. Somewhere around 1743, Mr. Franklin attended a lecture given by Professor Spencer of Edinburgh. Being a man of great imagination and enthused by Spencer's demonstrations, he was compelled to pursue this strange phenomenon with his own investigations and experiments.

He was a serious experimenter and immediately bought himself a generating machine and a Leyden Jar. The results of his experiments led him to believe that there were, indeed, two kinds of electricity, which he called positive and negative. He concluded that this force flowed from the positive to the negative.

Imagining that current flows from positive to negative must have surely been an effect of genius. It is now called *hole current*. One way of thinking of hole current is to imagine yourself suspended above a line of traffic backed up by a stop light. When the light turns green, cars move through the intersection and as they do, there is always a gap or space behind them. Looking down from

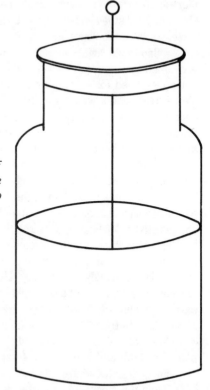

Fig. 2-2. The Leyden Jar is about the size of a canning jar, with a cork lid. The inside and outside of the jar is coated with tinfoil to about half way up. The glass in between is the insulator or dielectric.

above, you would notice this space, or hole, moves backward as the traffic moves forward. We normally think of electrons flowing from negative to positive. In actual practice—for example, a light bulb—it doesn't matter which way the current came; as long as it gets there, the lamp will still burn.

In 1752 Franklin flew his famous kite. By choice, or perhaps by luck, he picked a storm that was without lightning, for at about the same time a Russian scientist was killed by lightning while holding a rod up during a thunderstorm.

Franklin had gone about the experiment differently. At the top of his kite, he had fixed a stiff wire pointing up. The key was hung from a string on the other end of the kite string. His Leyden Jar was positioned close to the key. The falling rain began to moisten the kite string, and as water is a good conductor, the wet string began to conduct electricity down to the Leyden Jar. Sparks began to jump from the key to the jar. This proved to Franklin, that lightning and electricity were the same thing.

JAMES WATT

Although he conducted no experiments with magnetism or electricity, one pioneer must not be overlooked. James Watt was born January 19, 1736 in Greenock, a small village near Glasgow, Scotland. His grandfather was a professor of mathematics and taught surveying and navigation. His father, James, was a shipwright and a supplier of navigational instruments. It was generally presumed that young James would follow in his father's footsteps, so he was not apprenticed for a trade or attended a university. Even his attendance at elementary school was somewhat irregular because of his fragile health. The family-owned ship was lost in a storm along with the family fortune, so James Watt, at age 19 had to find a new career. He decided to become an instrument maker and set up a repair shop at the University of Glasgow in 1757.

Although relatively unschooled, he had a quick mind, and this atmosphere fostered much of his technical and scientific work. Here he met John Robison who turned James' attention to the steam engine. He had frequent discussions with Robison and another man named Joseph Black. Watt was fascinated by the characteristics of heat and its relation to steam.

The Newcomen steam engines were very popular at that time and one was to be used in a classroom demonstration. However, it failed to work and was sent to James Watt's repair shop. He finally got the machine running but was amazed to find that the boiler could not keep the engine running because of its extravagant use of steam. For several years he was troubled by this waste until 1765, still with the university, he created his first and most important invention: a separate container, or condenser for the Newcomen engine. He received a patent in 1769 and began to develop it commercially.

James Watt was interested in producing power to lift the burden of heavy labor from the working class. He received a patent for a steam locomotive in 1784. There is no doubt that his invention promoted the steam revolution. He died at Heathfield, England, August 19, 1819. In recognition of his contributions to science, the Second International Electrical Congress at a meeting in Paris in 1889 named a unit of power the Watt.

THE SOURCE OF ELECTRICITY

In Italy in 1780 Luigi Galvani (1737-1798) was conducting a class in anatomy at the University of Bologna. He was involved in dissecting a frog next to a generating machine that had been used in an earlier experiment when a spark passed between the frog and the machine. Galvani thought he had discovered an electrical source in animals. He continued his experiments and about eleven years later published the results.

A professor of physics at the University of Padua strongly disagreed with Galvani. Alessandro Volta (1745-1827) had a very different idea as to where electricity came from. A bit of a feud soon developed. Both men were well-respected, and each had his own followers. Volta believed that electricity came from the metals, or perhaps the difference between two metals. He began conducting his experiments and on March 20, 1800 sent a letter to the Royal Society of London describing his results.

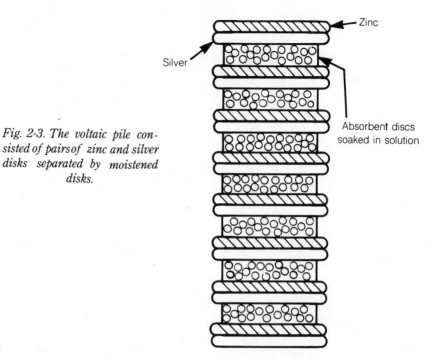

Fig. 2-3. The voltaic pile con-sisted of pairs of zinc and silver disks separated by moistened disks.

Volta had constructed a stack of pairs of discs, one zinc and one silver, with each pair separated by a leather or paper disc. These absorbent discs separating the two metals had been soaked in some kind of solution, salt water or perhaps vinegar or lemon juice. The stack began with a silver disc and ended with a ·zinc disc, two different kinds of metal. This was known as the voltaic pile (Fig. 2-3), sometimes referred to as an electric cell or battery.

There was some truth in both men's convictions. You need the two different kinds of metals as well as the chemicals. The chemicals in the frog's leg aided in the generation of the spark.

TRY IT OUT

You can duplicate Volta's experiment by using a stack of pennies and dimes separated by a piece of paper towel soaked in vinegar and water (Fig. 2-4). Start with an equal number of pennies and dimes; say, ten. Make sure they're clean for good conduction. Add a tablespoon of vinegar to a glass of water for the solution. If you began the stack with a penny, you should end with a dime. Begin with a penny, then add a small piece of paper towel soaked in the solution, then a dime, and another piece of wet paper towel. Repeat this with another penny, paper towel, a dime, more paper towel, etc. What you want is a stack of different metals separated from each other by a small piece of wet paper. A meter connected to each end of the battery, (penny on one end, dime on the other)

21

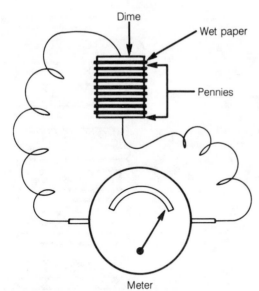

Fig. 2-4. A homemade voltaic pile or battery. Materials required: Ten pennies, ten dimes, paper towel, vinegar and water, and a dc voltmeter.

should show a brief deflection of the needle. Of course, the larger the stack, the greater the electrical force.

Electrical force, or pressure, is now measured in units called volts, after Alessandro Volta.

Without sacrificing a frog, you can do Galvani's experiment by substituting a lemon (Fig. 2-5). Roll the lemon on a table to make it juicy. Then cut the top off. A steel wire and a copper wire inserted opposite each other in the exposed area are the terminals. Just like the battery, one is positive and the other will be negative. Electrons flow from one terminal to the other because of the electrochemical action.

DISCOVERIES CONTINUE

On the 14th of June, 1736, a boy was born into a well-to-do family living in the southwest of France. Charles Augustin Coulomb (1736-1807) showed an early talent in astronomy and mathematics. His education continued into engineering and then later turned to physics. With an inquiring mind, he studied the effects of friction and cohesion. He then began to address the effects of shear as applied to bodies under torsional stress. Wouldn't a length of twine under some tension and then twisted slightly have some amount of torque when released? He found that by balancing an arm on a length of light fiber, he had a very delicate instrument that could be used to measure torque. This experiment led him to invent the first precision electrical instrument, the Torsion Balance

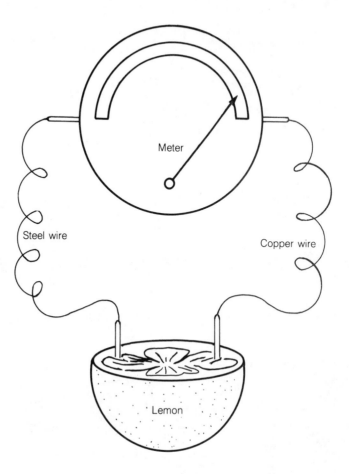

Fig. 2-5. The lemon battery. Materials needed: one softened lemon, a length of iron or steel wire, a copper wire, and a dc voltmeter.

(Fig. 2-6). It was a complicated apparatus but it could accurately measure the magnetic force in an electrical circuit.

By 1789 the French Revolution had began, and one by one the older institutions were overturned and reorganized. It became very dangerous for Coulomb to live in Paris. In 1793 he went to live at a small estate he owned in Blois, where he could continue his studies. Later when Napoleon came to power, Coulomb moved back to Paris and resumed his research until his death on August 23, 1806. This talented man and his contributions to electrical science were honored by the International Electrical Congress in Paris in 1881 when they named the unit of an electrical charge a *coulomb*.

In the winter between the years 1819 and 1820, a physics professor at the University of Copenhagen was conducting a class to show that electrical current could produce heat in a wire, a principle used by our toasters and electric heaters

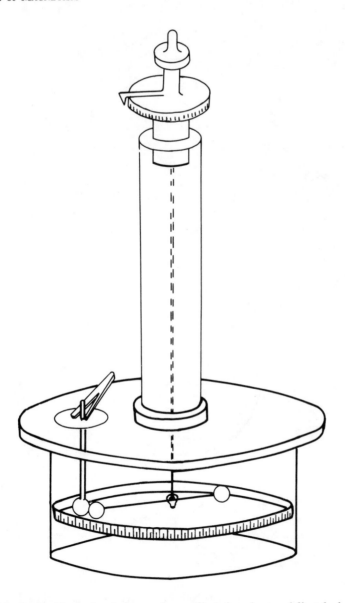

Fig. 2-6. Coulomb's torsion balance. A complicated and very delicately balanced instrument used to measure the forces of electrostatic charges.

today. Hans Christian Oersted (1777-1851) motioned for his class to come closer to the table to witness his demonstration. Professor Oersted then gave the signal for his assistant to connect the battery which would supply the current to heat the wire. By design or perhaps by chance, a compass had been left near the wire. The professor was surprised when he noticed that the needle of the compass suddenly came to life and pointed directly at the wire. He signaled the assistant to disconnect the battery. The compass needle swung back to its north-south direction. He again signaled for the battery to be connected. The compass needle immediately responded and pointed toward the wire (Fig. 2-7). It should be pointed out that magnetism and electricity were still regarded as two separate and distinct phenomena. Of course, the wire never had a chance to get hot, for the class had just witnessed their professor's discovery of the first evidence of a force connecting electricity and magnetism.

In a publication dated July 21st of that same year Oersted announced one of the greatest discoveries in the history of electrical science. "Experimenta

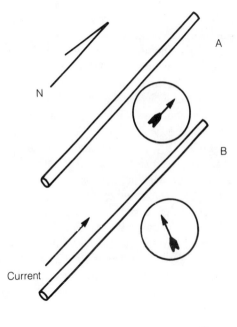

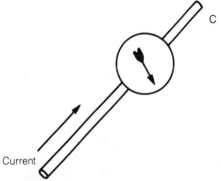

Fig. 2-7. Oersted's discovery of the magnetic effect of electric current. (A) With current off, compass points North. (B) With current on, needle points to wire. (C) Current is still flowing and with compass over wire, needle now points in opposite direction indicating a circular magnetic field around the wire.

Circa Effectuma Conflictus Electrici in Acum Magneticam" was a four-page document describing the results of his experiments. The significance of his discovery was immediately recognized and within a matter of a few weeks his pamphlet was reprinted into five different languages.

Oersted was introducing a new force that enjoys the company of an electrical current: *electromagnetism*. Hans Christian Oersted gave to the world the important discovery that electrical current does not flow alone in a wire but is enveloped in an invisible force field of magnetism.

Other than volts, the most common term used by electricians is amps. This term refers to Andre-Marie Ampere, born in Lyons, France the 22nd of January 1775. He grew up in a nearby village called Poleymieux, where his home is still a national museum.

His father, Jean-Jacques, was a well-to-do merchant who believed his son would be best educated by exposure to a large library, allowing the lad to educate himself in whatever areas his curiosity would lead him.

Ampere's sheltered childhood ended when he was fourteen with the outbreak of the French Revolution. During the ensuing turmoil, his father was called upon to assume a post as a judge. The position carried significant police powers, one of which was to order the execution of Joseph Chalier, a leader of the revolution. Consequently, when Lyons fell to the revolutionaries, Jean-Jacques Ampere was guillotined (November 23, 1793). We can only imagine the grief of an 18-year-old boy watching his father beheaded. Andre and his father had been very close and it was some time before he recovered from the shock, but eventually he met Julie Carron and with much enthusiasm pursued her till they were married on August 7, 1799. His emotions were to run from one extreme to the other. A son, Jean-Jacques, was born on August 12, 1800. The few years following were probably the happiest of his life. Tragedy struck again in the death of Julie on July 13th, 1803.

He had began his career as a mathematics teacher at Lyons and after Julie's death experienced another marriage which turned out to be a disaster. In spite of this, by 1820 he had achieved a reputation as a mathematician and somewhat of a chemist.

On September 4th, 1820, a man by the name of Francois Argo announced Hans Christian Oersted's discovery of the relationship between magnetism and electricity at a meeting of the Academic des Sciences. Ampere was fascinated and immediately accepted Oersted's discovery as fact.

All of his enthusiasm was immediately turned toward the study of this new phenomenon and just two weeks later was able to give a presentation on the subject to the academy. On the 25th of that same month he read his second paper to the academy, and by the 9th of October had developed the science of electrodynamics.

He explained later that his thoughts had gone from the existence of electromagnetism to the notion that a current orbiting through a helix, or coil, would create a magnet. This was based on Oersted's discovery suggesting that a current-carrying wire might have some effect on another current-carrying wire. His paper on the 25th also mentioned the mutual attraction and repulsion of two coils.

Ampere was offering a new theory which presented magnetism as electricity in motion. His experiments had shown that one electric current would be attracted to another parallel electric current if they were moving in the same direction; however, if one was reversed, they would repel each other. This effect was completely different from the one caused by static electricity (Fig. 2-8). Ampere developed a deep religious faith, no doubt strengthened by a life filled with one tragedy, or catastrophe, after another. He had had great hopes for his son, but Jean-Jacques wasted twenty years as one of the escorts to one of France's great beauties, Mme. Recamier. His daughter, Albine, from his ill-fated second marriage, had married an army officer who was almost a maniac as well as a complete drunkard. And money was always a problem for Ampere.

Andre-Marie Ampere died alone in Marseilles on the 10th of June, 1836. Almost fifty years later the First International Electrical Congress, in 1881, defined a unit of electrical current the ampere.

Anything that moves, or flows, encounters some amount of resistance. A rule that reflects this phenomenon was developed by a German professor named Georg Simon Ohm (actually he was christened Johann), born at Erlangen, Bavaria

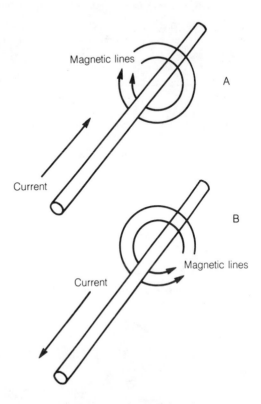

Fig. 2-8. Magnetic forces between two wires. (A) Direction of magnetic field when current flowing left. (B) Magnetic field reversed when current flowing right. The right hand rule: when your right hand is curved down over wire, extended thumb indicates direction of current while fingers indicate direction of magnetic field. With hand curved up around wire, opposite directions are indicated.

on March the 16th, 1789. Born into a poor family he experienced the frustrations of poverty. His father, Johann Wolfgang Ohm, was a master locksmith and understood the importance of a formal education. He tutored Georg on basic mathematics, physics, and chemistry. By his early twenties Georg had earned a doctorate degree in philosophy from the University of Erlangen where he became a teacher of mathematics. However, the Napoleonic war disrupted his situation and he eventually obtained a teaching position in Cologne where he was to remain for over nine years.

Enthused by the discoveries of Oersted and Ampere and fortified by a well-equipped lab at the school, he began conducting experiments in electricity and magnetism. Ohm studied the conducting abilities of different metals of varying lengths. He included in his studies the cross-section of wires and the thickness of conductors, and the effect this had on current-carrying capabilities.

His first paper, published in 1825, was concerned with the loss of magnetic force due to the length of a current-carrying conductor. He discovered that the conduction of electricity through wires varied not only with the metal composition of the wires but also with the dimensions of the conductor. His discovery means that a larger size of wire is used in your home for an electric dryer than is used for a light fixture. Because of the diligence of his investigations into the effects of Ampere's current supplied by Volta's energy force, he invented a precise method of measuring the resistance to this electrical current (Fig. 2-9).

Ohm was an excellent teacher and in 1841 the Royal Society of London presented him with their highest award, the Copely Medal. He spent his final

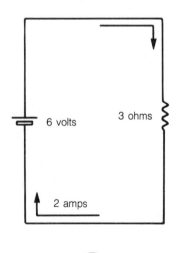

Fig. 2-9. Ohms law. Cover the letter of the factor you want to solve for with your finger and calculate the two remaining. The 6 volt supply with a 3 ohm resistor allows a current of 2 amps to flow.

years at a respected post at the University until he died in 1854 at age 65. In 1862, a year after it was appointed, the British Association Committee of Standards of Electrical Resistance named the standard unit of resistance the Ohmad (later shortened to the Ohm).

MICHAEL FARADAY

In 1881, the First International Electrical Congress called the standard unit of capacitance the *farad*, for Michael Faraday. A capacitor consists of two metal plates separated by an insulator. The capacitance, or size, of a capacitor is measured in farads. A small honor, indeed for the 19th century genius who laid the foundation for modern physics.

Three centuries earlier during the bloody Irish rebellion, the Faraday family had escaped to the west coast of England, later moving eastward and settling in a small village in Yorkshire. James Faraday was a blacksmith and did reasonably well until the depression that followed the French Revolution in 1789. Hoping that business would be better in the city, James Faraday moved his family to the outskirts of London where Michael was born in late September in 1791. At age 14, Michael became apprenticed to a bookbinder. With a large amount of reading material available and a very curious mind, he was able to educate himself. Michael was attracted to scientific publications and more particularly to articles on chemistry and electricity. It was only natural that he began his own experiments with Leyden Jars and frictional generating machines.

In 1810 he became a member of The City Philosophical Society, a group of men interested in science and dedicated to self-improvement. They met at the house of John Tatum each Wednesday night. Tatum would give lectures, including demonstrations, every other Wednesday on natural philosophy, which was their term for natural science.

Not much was known about electricity or magnetism at that time and various theories were presented at these meetings. Many of the theories Michael Faraday was quick to disagree with. He was an original thinker, not swayed by popular opinion. His association with the society provided him with the self-confidence and poise he needed to further his scientific career.

In the spring of 1813 Faraday obtained a position as an assistant to the director of the Royal Institution, Sir Humphry Davy. The Royal Institution of Great Britain, usually just called the Royal Institution, was founded by a group of men in 1799 led by "Count Rumford." It turns out "Count Rumford" was an American politician and scientist by the name of Benjamin Thompson (1753-1814). Its purpose was to encourage the education of science and then to apply it for the usefulness of the common man.

Davy had married a rich widow in 1812 and in 1813 elected to spend two years traveling France, Italy, Switzerland, and Germany. Faraday, as his assistant, accompanied Sir Humphry and Lady Davy. They concluded their tour in Rome and returned to England in the winter of 1815. Faraday, whose travels had previously been limited to the immediate area of London, was now well-

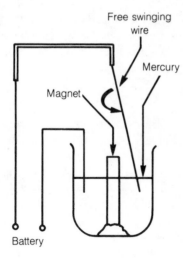

Fig. 2-10. Faraday's electromagnetic motor. When a battery was connected to the wires the current developed a magnetic field causing the free hanging wire to rotate around the magnet.

traveled and intellectually established. He had developed conversational abilities in French and Italian as well as establishing relationships with famous scientists of these countries. He had the right stuff at the right time. Late in 1820, the news of Oersted's discovery came to the attention of the Royal Institution. Davy, assisted by Faraday, immediately began to repeat Oersted's experiments. Faraday was unable to participate in all of the experiments due to other work and his romance, and later marriage, to Sara Barnard. In late 1821, he came up with an ingenious apparatus to prove that electrical energy could be converted to mechanical energy through magnetism (Fig. 2-10).

He positioned a magnet with one end exposed in a cup of mercury. A wire was suspended from above until it touched the mercury. A battery was connected in series between the mercury and the other end of the wire. The free end of the wire immediately began to whirl around the magnet. Michael Faraday had invented the electromagnetic motor. He quickly published a paper describing the results of his discovery, the significance of which was immediately recognized as a scientific breakthrough. He was instantly hailed as one of Europe's top scientists.

Davy did not appreciate being outshone by his assistant and when Faraday was considered for membership in the Royal Society only one vote, Davy's, was against him. It was of little consequence because he was fully accepted in 1824.

It occurred to Faraday that if electricity could produce magnetism couldn't magnetism produce electricity? Due to the responsibility of other assignments he was unable to pursue this reasoning until late 1831, when he discovered the principle which enables a utility company to present you with a monthly electric bill. (Remember now, he didn't invent the rates, just the principle!)

Michael Faraday made two important discoveries in the fall of that year. The first was based on an experiment he conducted with the Faraday ring (Fig. 2-11). Using an iron ring about six inches in diameter, he constructed two helixes, or coils, one on each side of the ring. Between one coil of wire he connected a galvanometer, which could indicate very small amounts of current or voltage. To the other coil he connected a battery while monitoring the deflection of the needle on the meter. When the final connection was made to the battery, the needle on the meter spun around several times and then stopped. When the battery was disconnected, the needle immediately spun in the opposite direction.

He was quick to realize that the current only flowed through the meter when magnetism was developed as he connected or disconnected the battery. Then couldn't he move the magnet and also deflect the needle (Fig. 2-12)? He realized that it was the change of the magnetic field of the first coil that deflected the needle connected to the second coil. What he had constructed was basically a transformer.

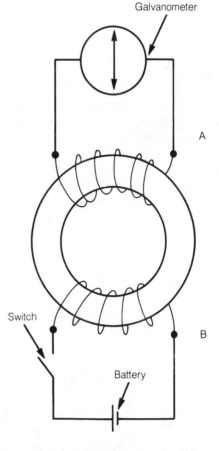

Fig. 2-11. Faraday's ring. The opening and closing of the switch on side A caused a rise and fall of the magnetic field of the ring which induced a current in the coil on side B that caused a deflection of the needle of the galvanometer.

Galvanometer

A

Switch

B

Battery

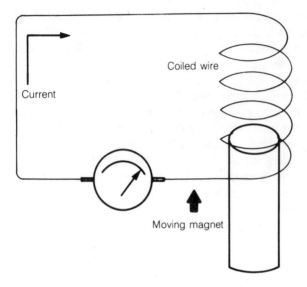

Fig. 2-12. Converting magnetism to electricity. When a magnet abruptly travels inside a coil of wire an electric current is developed in the wire.

If he moved the magnetic field inside a coil it might produce similar deflection of the needle. Connecting a galvanometer across the coil, he found that he could cause a deflection of the needle when he abruptly inserted and withdrew a permanent magnet in the coil. This action of one coil inducing an effect on another coil he called induction. This is an important term now used to describe results of a magnetic field applied to another circuit. This basic law of physics that pertains specifically to the effect of magnetism is what governs the operational capabilities of electric motors and generators.

Michael Faraday conducted many experiments dealing with the ability of electromagnetic waves to pass through different materials. One such experiment involved the use of a polarized light passing through a heavy pane of glass. Normally the light would pass through the glass unaffected; however, when he exposed the glass to a strong magnetic field, the beam was altered. Logic led him to believe that somehow the glass was affected by the electromagnetic waves.

Next, he carefully balanced a piece of glass between the poles of a strong electromagnet. The glass tried to turn opposite, or across, the magnetic lines of force. A piece of magnetic material would want to align itself with the lines of force. Faraday called this new phenomena *diamagnetism.*

The results of many experiments proved that most things, or materials, were diamagnetic. However, a few of the materials he tested did try to align themselves along the lines of the magnetic field. These materials he called *paramagnetic.*

By 1862, he was trying to find a connection between the magnetic waves and the wave length of light. Light waves and magnetic waves somehow should affect one another, but his experiments failed to produce the results he expected, primarily because his detecting instruments were not sensitive enough to indicate any results.

Faraday was a gentle man and although not wealthy could often be found helping needy families with his own money. He became a popular speaker and developed a special series of scientific lectures for children delivered each Christmas which still continue and are shown on television. Michael Faraday died peacefully on the 25th of August, 1867.

Magnetic Materials

A magnet is a metal, normally iron, that either attracts or repels a certain group of other metals. There are only two types of magnets: permanent magnets and electromagnets.

Think of magnetic materials as either hard or soft. Soft materials are used in devices where a change in the magnetic field is necessary in the operation of the device, sometimes very rapidly. These are electromagnets. An electromagnet can be activated by a switch; the magnetic field can be turned on and off. This effect is necessary in the operation of solenoids, relays, electric brakes and clutches, as well as ac generators and transformers.

HARD AND SOFT MAGNETIC MATERIALS

Hard materials are used to supply a fixed magnetic field that acts alone, such as the magnet that holds a note on a refrigerator, a large magnetic separator used in mining, or the magnet in the back of a speaker in a radio (Fig. 3-1). These are permanent magnets.

Earth's natural resources offer magnetite, or the loadstone, which is a permanent magnet. Magnetite comes to us in the form of minerals and has developed into a very important ore in the manufacturing of iron. It can be found in black crystal form as well as in laminations in rocks or in granular forms such as beach sand. Magnetite can also be found in meteorites and in a natural abrasive called emery.

The world's largest concentrations of deposits of magnetic material are found

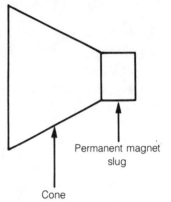

Fig. 3-1. Schematic symbol of a loudspeaker.

Permanent magnet slug

Cone

in northern Sweden. The strongest deposits are found in the Harz Mountains in Siberia and on the island of Elbe between Corsica and Italy. Magnetite is nature's permanent magnet. Artificial permanent magnets are made by magnetizing a piece of iron or steel such as a horseshoe or bar magnet (Fig. 3-2). The needle of a compass is a permanent magnet.

Unlike a permanent magnet, an electromagnet is a temporary magnet, usually a coil of wire wound around a piece of iron. This device develops a

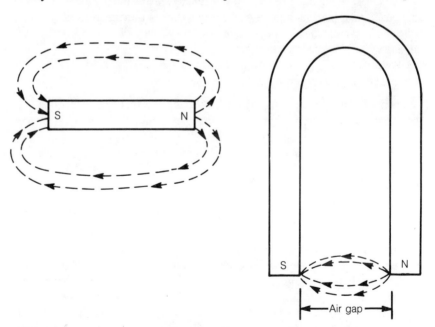

Fig. 3-2. Bar magnet and horseshoe magnet. The horseshoe magnet is a bar magnet bent double to reduce the air gap between the poles, which increases the strength by concentrating the magnetic field.

35

magnetic field as long as an electric current is flowing through the coil. The solenoid that engages the starter on a car is a good example. Larger versions can be observed as cranes loading and unloading scrap iron.

PERMEABILITY

Magnetism affects all materials to a lesser or greater degree. This is called *permeability*. The dictionary tells us that permeate can mean to saturate, penetrate, or perhaps a passage. A French engineer, Henry Darcy (1803-1858), was the first to analyze permeability as the flow of water through a sand-filled pipe. The capacity of a body of a porous sediment to transmit fluids depends on the closeness or connections of spaces between the pores along with the size of the area of the body.

Analyzing electrical circuits can often be simplified by thinking of them as a flow of a fluid in a network of plumbing. It follows that the ability of a substance to pass or conduct the lines of a magnetic force is called the permeability of that material. Picture in your mind an invisible force approaching an object, and then passing through this object. Relative permeability compares the ability of a material, or substance, to conduct a magnetic force and to the ability of empty space to conduct the same magnetic force. Relative permeability is a pure number; that is, there are no units, just number values.

To measure anything requires a reference or a starting point. In this case, air is used as a beginning point and permeability is a measurement from this point. For example, an iron rod can have a permeability of four or five thousand depending on the grade of the iron.

FERROMAGNETIC MATERIALS

Different materials produce different phenomena that can be classified into three basic groups: ferromagnetic, paramagnetic, and diamagnetic. The most strongly magnetic materials are called ferromagnetic. This group includes iron, steel, nickel, and cobalt along with some commercially-made alloys such as alnico and permalloy. Iron was known in prehistoric times. Meteors probably provided early man with his first samples of iron. Ancient writings have referred to the "metal from heaven."

The beginning of the Iron Age was around 1200 B.C. (the Iron Age is the last of three archaeological stages known as the Stone Age, Bronze Age, and the Iron Age). When and where it was first smelted is unknown. The process was probably discovered accidentally when sources of copper and tin to make bronze may have become scarce. Somehow man had learned to extract this element from its ores. This silvery-white metal has the chemical symbol Fe which comes from the latin word for iron, *ferrum*. It has the atomic number 26 in the periodic table and makes up about 5 percent of the elements in the Earth's crust.

The atomic number is the number of protons or electrons in the atom of each element. In the simplest atom, hydrogen, the atomic number is 1. This

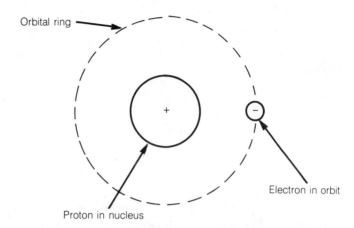

Orbital ring

Electron in orbit

Proton in nucleus

Fig. 3-3. One electron spinning in orbit around one proton in an atom of hydrogen.

means the nucleus has one proton balanced by one orbiting electron (Fig. 3-3). The periodic table is a chart showing the arrangement of the chemical elements according to their atomic numbers (Table 3-1). The higher the atomic number, the more orbiting electrons the element has.

The elements that deal with the area of electricity and magnetism are called transition elements, which are the metals. These elements are arranged in such a manner that they progressively add an electron to the shell of the previous element of a lower energy. The shells, or orbital rings of the planetary electrons, are also called energy levels (Fig. 3-4). These transition elements are the ones that add electrons to an inner shell after some electrons have moved to positions

Table 3-1. Periodic Table.

CONVENTIONAL REPRESENTATION OF PERIODIC TABLE

1a																	0
1 H	2a											3a	4a	5a	6a	7a	2 He
3 Li	4 Be											5 B	6 C	7 N	8 O	9 F	10 Ne
11 Na	12 Mg	3b	4b	5b	6b	7b		8		1b	2b	13 Al	14 Si	15 P	16 S	17 Cl	18 Ar
19 K	20 Ca	21 Sc	22 Ti	23 V	24 Cr	25 Mn	26 Fe	27 Co	28 Ni	29 Cu	30 Zn	31 Ga	32 Ge	33 As	34 Se	35 Br	36 Kr
37 Rb	38 Sr	39 Y	40 Zr	41 Nb	42 Mo	43 Tc	44 Ru	45 Rh	46 Pd	47 Ag	48 Cd	49 In	50 Sn	51 Sb	52 Te	53 I	54 Xe
55 Cs	56 Ba	57 La	72 Hf	73 Ta	74 W	75 Re	76 Os	77 Ir	78 Pt	79 Au	80 Hg	81 Tl	82 Pb	83 Bi	84 Po	85 At	86 Rn
87 Fr	88 Ra	89 Ac															

LANTHANIDES

58 Ce	59 Pr	60 Nd	61 Pm	62 Sm	63 Eu	64 Gd	65 Tb	66 Dy	67 Ho	68 Er	69 Tm	70 Yb	71 Lu

ACTINIDES

90 Th	91 Pa	92 U	93 Np	94 Pu	95 Am	96 Cm	97 Bk	98 Cf	99 Es	100 Fm	101 Mv	102 No	103 Lw

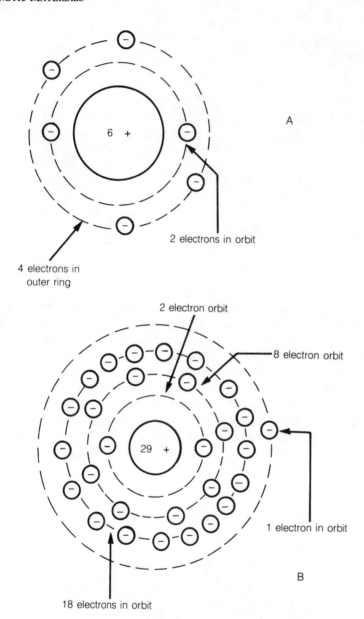

A

2 electrons in orbit

4 electrons in
outer ring

2 electron orbit

8 electron orbit

1 electron in orbit

18 electrons in orbit

B

Fig. 3-4. Atomic structure. (A) Carbon atom with 6 protons in the nucleus and 6 orbiting electrons. (B) Copper atom with 29 protons and 29 orbiting electrons.

in an outer shell. This happens because the energy levels of the shells overlap. Transition elements have one or two electrons in the outermost shell of their atoms. They are usually hard metals with high melting and boiling points. They generally lose rather than gain electrons. It also happens that iron is the central atom in *heme* which turns out to be the oxygen-carrying part of hemoglobin found in the blood.

Iron at room temperature has the strong magnetic characteristics of ferromagnetism: once magnetized, it can exhibit this magnetic phenomenon without the applied magnetic field. However, when it is heated to a temperature of 768 degrees centigrade it exhibits a paramagnetic phenomenon which is a weaker attraction to a magnetic field. Iron does conduct electricity but is generally considered to be a poor conductor when compared to copper or aluminum. Iron's magnetic properties can be improved by mixing in other elements, usually metals, forming alloys. Probably the most important alloy is steel which can have up to about 2 percent carbon. Add one to five percent silicon, and you have an alloy that is hard and highly magnetic.

The iron-silicon alloys used in magnetic circuits are known as silicon steels and have high values of permeability. The product is generally produced in flat-rolled or sheet form and classified as grain-oriented or nonoriented. Grain-oriented steels are processed in a special way that tend to align the magnetic properties in a preferred direction. They contain abut 3.25 percent silicon and are used in very high-efficiency transformers and turbine generators of the kind used by utility companies for power distribution.

The nonoriented steels that contain 0.5 to 1.5 percent silicon, the low-silicon class, are used in the rotors and stators of motors as well as generators. One percent silicon steels are also used in relays and small transformers. Intermediate-silicon steels of about 2.5 to 3.5 percent silicon are used in small- to medium-size motors, generators and transformers. High-silicon steels of 3.75 to 5 percent silicon are used in communication equipment as well as power transformers and high efficiency motors and generators.

Iron-aluminum alloys containing about 12 to 16 percent aluminum are considered magnetically soft materials and are used in ac magnetic circuits. Iron-chromium-aluminum alloys have an electrical resistance that make them useful in manufacturing heating elements used in electric heaters.

Iron-nickel alloys can have from 40 to 60 percent nickel with the highest saturation value achieved at about 50 percent. They have high permeability with low magnetic losses. Iron-nickel alloys are used in the manufacture of audio transformers, magnetic amplifiers, and shields as well as choke coils and relays. Alloys containing about 30 percent nickel can be used in magnetic circuits when it is beneficial to compensate for changes that occur due to temperature changes. The permeability tends to decrease at a predictable rate as the temperature is increased. Nickel, Ni, has atomic number 28 in the periodic table.

Iron-cobalt alloys with up to 65 percent cobalt have a higher saturated value than pure iron. An alloy containing 49 percent cobalt and 2 percent vanadium can be used in telephone diaphragms, magnetic yokes, and ultrasonic equipment.

Cobalt resembles iron and nickel and makes up about 0.001 percent of the igneous rocks of Earth's crust. Cobalt (Co) has the atomic number 27 in the periodic table. Iron-base and cobalt-base alloys with 6 to 65 percent cobalt along

Table 3-2. Composition of Alloys of the Alnico Type.

Name	Composition, percent of weight, balance is iron
Alnico 2	12.5 Co, 17 Ni, 10 Al, 6 Cu
Alnico 5 (Ticonal)	24 Co, 14 Ni, 8 Al, 3 Cu
Alnico 5 DG	24 Co, 14 Ni, 8 Al, 3 Cu
Alnico 6	24 Co, 16 Ni, 8 Al, 3 Cu, 2 Ti
Alnico 9	35 Co, 15 Ni, 7 Al, 5 Ti
Alcomax II	22 Co, 11.5 Ni, 8 Al, 3 Cu
Alcomax III	24 Co, 14 Ni, 8 Al, 3 Cu, 1 Nb

with some chromium and tungsten are very hard and resist corrosion. A good magnetic steel can contain about 35 percent cobalt, 6 percent chromium, and about 4 percent tungsten. Cobalt can be found as a component in some electrically resistant alloys such as beryllium copper and cathode filaments.

Cobalt comes from the German word *kobold* which means a malicious demon or goblin that lives underground. It is never found in its pure state but is usually bonded to sulfur and arsenic, which probably contributes to its expense in mining. Cobalt can also be found in meteorites. It is the center of the three metals making up the iron group in the periodic table: iron, cobalt, and nickel. It is a relatively expensive metal. Cobalt when mixed with iron creates an alloy with special magnetic properties. Hyperco, for example, is used for the base for strong electromagnets.

Alloys of cobalt, titanium, aluminum, and nickel, such as alnico, are used in making permanent magnets. Alnico is the name of a series of alloys used in manufacturing permanent magnets (Table 3-2), Alnico 5 being the most commonly used. These magnetic materials are usually prepared by melting, casting into the desired shape, reheating to about 1300 degrees Centigrade, and then slowly cooling inside a magnetic field.

These ferromagnetic materials can become strongly magnetized in the same direction as the magnetizing force. They generally have high values of permeability, from 50 to 5,000. Permalloy has a permeability of 100,000. Permalloy contains about 80 percent nickel and 20 percent iron with optimum results obtained with 78.5 percent nickel. It is highly magnetic in a weak magnetic field but loses its magnetism faster when the field is removed than other metals. This effect makes permalloy an important material in some electrical conductors.

Cobalt is produced in Zaire, Russia,and Australia. It turns out that cobalt is essential to life itself. Pernicious anemia can develop when the red blood cells fail to develop and mature normally and as late as 1925 this disease was usually fatal. It is caused by a deficiency of a cobalt compound called Vitamin B-12.

PARAMAGNETISM

Paramagnetism is the effect of a weak magnetic attraction in which materials tend to set themselves along the lines of a magnetic force. The permeability of paramagnetic materials is slightly more than one. They can become weakly magnetized in the same direction as the magnetizing force. Some of the materials include aluminum, platinum, manganese, and chromium.

DIAMAGNETISM

Diamagnetism is a magnetic characteristic of materials that tend to line up at right angles to a magnetic field and even partly repel the magnetic field they are in. Diamagnetic materials include copper, silver, gold, mercury, zinc, antimony, and bismuth. Their permeability is less than 1. Each one can be weakly magnetized; however, it will be in the opposite direction from the original magnetizing field.

Antimony is a brittle, silver-white, metallic chemical element used to harden alloys. Bismuth is a greyish-white metallic element used mostly in making alloys that melt at low temperatures. In the periodic table, this shiny metal belongs to the same group as arsenic. It was apparently unknown in ancient times but its existence was known to the Europeans by the middle ages. Bismuth salts are used in the cosmetics and pharmaceutical industries.

Michael Faraday in 1845 originated the term diamagnetism to explain the results, or effects, of experiments he conducted with a piece of glass suspended between the poles of a powerful horseshoe electromagnet. After many tests he found that, indeed, all things are diamagnetic.

The powerful outside magnetic field slows down or speeds up the electrons orbiting in atoms of a material in such a manner that will oppose the action of the outside magnetic field. This diamagnetic action can be hidden by a weak magnetic attraction which would be called paramagnetic or a strong magnetic attraction called ferromagnetic.

FERRITES

Nonmetallic materials can also have the ferromagnetic properties of iron. These materials are called *ferrites*. Ferrites are a ceramic material with very high permeability, similar to iron, in an area of 50 to 3000. Unlike iron, a conductor, the ferrite is an insulator. A ferrite core, normally adjustable, is used in the coils of rf transformers because it is more efficient than iron at this high of frequencies.

Ferrite cores can be used in small coils and transformers up to about 20 megahertz. Their high permeability allows the physical size of the transformer to be very small. However, ferrites are easily saturated and are not used in power transformers. Ferrite beads strung on a bare wire can concentrate the magnetic field of the current in the wire in such a way that current for a specific undesirable radio frequency is reduced. This effect allows them to take the place of a coil as a simple and economical rf choke.

One of the earliest attempts to explain the phenomenon of magnetism and magnetic materials was made by Ampere, well over a century ago, who suggested that the magnetization of a material is caused by the orientation of the circulating currents of the molecules of that material.

Although iron usually comes to mind when you think of a magnetic material, all materials are magnetic to some degree.

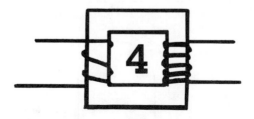

Electromagnetism in the Early Years

Based on Faraday's experiments, other scientists began to construct hand-cranked generators. These generators operated by moving a coil back and forth in front of a permanent magnet or by moving a magnet inside a coil in a similar manner. Today's generators use rotary motion between the coil and the magnet instead of the back and forth movement in a straight line.

Credit for this transition should be given to a French instrument builder named Hippolyte Pixii. In 1832, Pixii built the first apparatus that had fixed coils wound on a u-shaped iron rod. Above the rod a horseshoe magnet was attached to a shaft driven by a hand crank (Fig. 4-1). When the handle was turned, the rotating magnet produced an alternating electric current in the coils. Pixii's machine, which he created after reading papers on Faraday's work, was the forerunner to the same generators we use today. However, at that time there was no use for an alternating current.

It was Ampere that suggested an improvement to Pixii's machine: a cam-operated switching design that would reverse the alternations and produce a more usable current very near that of the battery. This innovation developed into the commutator, a mechanical switch that maintains current in one direction. This is the action of the brushes rubbing on the bars of the armature of a generator today.

JOSEPH HENRY AND ELECTROMAGNETIC INDUCTION

Another man who became famous for his discoveries in electromagnetism

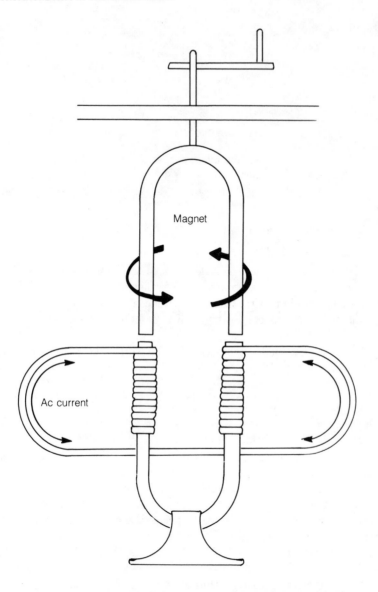

Fig. 4-1. Pixii's generator. The horseshoe magnet was rotated over the u-shaped armature creating an alternating current in the exposed armature leads.

was the American physicist Joseph Henry (1797-1878). He was actually the first to discover the principles of electromagnetic induction. Henry was born at Albany, New York, the middle of December, 1797, the son of a minister whose income afforded a limited education for young Joseph. He began to develop a career as an actor in his teens, but a chance reading of a book on science fired his imagination and he vowed to devote the rest of his life to the pursuit of scientific knowledge.

In 1819 he enrolled in the Albany Academy where he later taught mathematics and physics. It was during his years at Albany that he began conducting his experiments in electromagnetism.

One of his most important works was in improving the electromagnet. He wound the electromagnets he used in teaching his physics classes. These devices were developing unheard-of lifting power. A Yale College professor, Benjamin Silliman, asked Henry to make a lifting magnet for Yale's laboratory as strong as he could make it. The horseshoe magnet was not large by today's standards— about 12 inches tall—but it was the most powerful electromagnet of its time and could easily lift 2,000 pounds. Henry was later able to build one that lifted 3,600 pounds.

In his experiments he had also discovered a new phenomenon of electromagnetic reaction. He noticed when he connected a wire about 30 or 40 feet long across a battery he might see a small spark, but when the connection was broken a large spark occurred. He further observed that the effect was greater if he coiled the wire. Henry theorized that the wire had become charged with electricity that reacted within itself as the connection is broken. Now, this is called *self-induction.*

Joseph Henry failed to publish much of his work and received little historical credit. However, in 1846 he accepted the position as the first secretary and director of the Smithsonian Institution. In respect for his scientific achievements a unit of measurement of mutual inductance is called the *henry.* One henry of induction is produced when the current of one coil is alternated at one amp per second developing a magnetic field which induces one volt in a second coil.

GAUSS AND WEBER

The measurement of the strength of a particular area in a magnetic field, the flux density, is in units called *gauss.* This term comes from the German mathematician Karl Friedrich Gauss (1777-1855), probably one of the greatest mathematicians of all time. To begin with, he was a child prodigy. Karl Friedrich was born April 30th, 1777, to Gebhard and Dorothea Gauss. Gebhard was harsh and domineering. Uneducated, he provided a meager sustenance for his family as a bricklayer and laborer. Dorothea was probably able to balance this relationship somewhat. She was intelligent and had a strong character complemented with a humorous disposition.

At seven years of age, Karl was enrolled in school and by 10 was astounding his schoolmaster with his genius in arithmetic. By late 1795, Karl was admitted to the University of Gottingen as a student of mathematics. He returned to Brunswick in 1798 to continue his work in mathematics and astronomy. In 1805 he was appointed director of the University of Gottingen observatory which he would be associated with the rest of his life.

Degaussers used by Navy

On October 24th, Wilhelm Edward Weber (1804-1891) was born in Wittenberg near Berlin. During the Napoleonic wars the city was under attack by artillery and the Weber home was burned. Michael Weber, a theology professor at the University of Wittenberg, moved his family a few miles south to the city of Halle. Wilhelm was raised in a scholastic atmosphere and conducted scientific experiments at an early age. At 16 he entered the University of Halle and later, after completing his studies, was awarded a position as an assistant professor there.

His doctoral dissertation was on the theory of organ pipes. Weber was fascinated by the science of sounds, the field of acoustics. In 1831 he became a professor of physics at Gottingen. Here Gauss and Weber worked together on many experiments and by 1833 they had developed the electromagnetic telegraph. They were sending coded messages to each other over a two-wire circuit from the Gottingen observatory to another station about a mile and a half away.

Local politics separated the two scientists in 1837, when 18-year-old Queen Victoria ascended to the throne of England. Germany was made up of a group of kingdoms and Victoria's uncle, Ernst August, was now the King of Hannover. He did away with the constitution and demanded an oath of allegiance.

Weber and some other professors protested and were fired. Gauss did not protest and was retained. However, without Weber's help he soon abandoned his research in the physical sense. He continued to mathematically formulate the experiments of other scientists. He was a mathematician with astronomy dominating most of his life. Gauss made many important contributions in the field of electromagnetism and the ability to measure the strength of the magnetic field. He died peacefully in February 1855.

After being fired, Weber drifted for a half dozen years. Then in 1843, through the aid of friends, he obtained a position at the University of Leipzig where he collaborated with physicist Gustav Fechner. From this association Weber developed the theory that electromagnetic phenomena were the effect of forces exerted by electric charges in motion as well as stationary. His law of electrical force, published in 1846, was one of the first that dealt with electron theory and the speed of electric charges.

In 1856 Weber and Rudolph Kohlrausch (1809-1858) calculated the discharge between the coatings of a Leyden jar through the coil of a galvanometer. The velocity they obtained was 310,740,000 meters per second. The speed of light had already been measured to be from 314,000,000 m/s to 298,360,000 m/s by 1850. Based on these observations, G. R. Kirchhoff (1824-1887) stated in 1857 that electricity did indeed travel at the speed of light.

In honor of Wilhelm Weber the large unit used to measure magnetic flux (the lines in a magnetic field) is called the weber.

JAMES MAXWELL

Another great physicist of the 1800's was the Scottish scientist James Clerk

making his own toys. In school he displayed a quick mind with an incredible memory. He enjoyed poetry as well as mathematics and the study of languages.

In 1877 Nikola entered Polytechnic Institute at Graz. In his second year a situation occurred that strongly affected his future. The professor was demonstrating a machine that could operate as a dc motor or generator. Nikola asked whether the excessive sparking at the brushes could be eliminated (Fig. 4-3). His professor ridiculed his idea and embarrassed him in front of his fellow students.

Nikola became determined to solve the problem of a brushless motor. Four years passed, then, in 1882 in the city park at Budapest, the entire concept came to him in an instant. By using out-of-phase alternating currents flowing through coils he could develop a rotating magnetic field (Fig. 4-4). This would be the operating principle for his new motor. However, it was some time before he could develop it commercially.

In 1884 Nikola Tesla landed in New York City. At that time there were two types of electrical distribution in the United States. Thomas Edison's direct current supply was restricted to short distances with a powerhouse necessary for each square mile. Serving a downtown area with many lights required huge copper conductors to handle the higher current. The second source

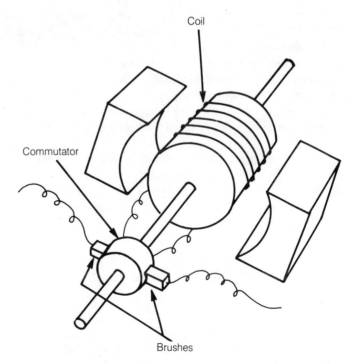

Coil

Commutator

Brushes

Fig. 4-3. The dc motor and generator are basically the same machine. It was the arcing of the brushes that disturbed Tesla.

49

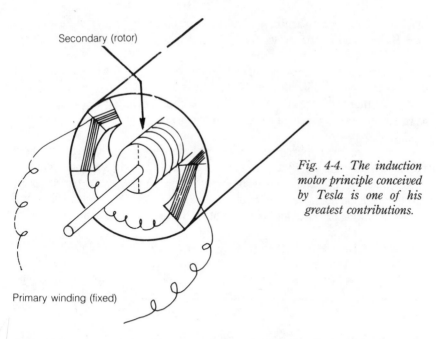

Secondary (rotor)

Primary winding (fixed)

Fig. 4-4. The induction motor principle conceived by Tesla is one of his greatest contributions.

of electricity was an alternating current being promoted by George Westinghouse (1846-1914), founder of the Westinghouse Electric Company. This alternating current had a distinct advantage: the useful voltage developed could be raised or lowered by a transformer, while direct current could not.

This enabled a generating plant to transform the electricity to a high voltage and low current for distribution along a power line. Then, through another transformer, the voltage was lowered to a desired level at the destination. This meant that the low currents could use much smaller wires to transmit electricity over long distances.

Tesla worked at the Edison Machine Works in New Jersey where he tried to interest his employer in his new induction motor and his ideas on alternating current. Thomas Edison would hear none of it. Tesla left the company in 1885. Three years later he established his own laboratory in New York. Here he developed his motor system.

He used two types of rotors for his motors. One employed iron poles without windings. This rotor developed very little torque; however, when at operating speed, it became synchronized with the rotating field of the stator. This was the first synchronous motor.

The other rotor used windings and developed high starting torque but ran at a lower speed. This lower speed was just enough to induce energy, by transformer action, from the field windings to the rotor. This was the first induction motor. Tesla received three patents for his dream in 1888.

Nikola Tesla went on to develop and demonstrate a complete polyphase (currents having two or more phases) alternating current system that included the generator, transformers, and the distribution system to a motor and several lights.

The results of his work soon attracted the attention of George Westinghouse and on July 7, 1888, Westinghouse paid one million dollars, plus a royalty for each motor sold, for the patent rights. He also hired Tesla himself, but there was a problem: Tesla's induction motor was polyphase while existing services were single phase.

Tesla next invented the split phase motor which allowed the single phase service to do the work of two phases. The motor performed poorly because of the frequencies available at that time. The cycles per second of the current ranged from 16 to 133 cycles. Tesla pushed for a standard and established the 60 cycle frequency used today. Then his motor worked fine.

In 1893, Westinghouse won the bid to light the grounds at the Chicago Exposition. Here, in celebrating the 400th anniversary of the discovery of America, was a great opportunity to publicize the Tesla system. They put on a dazzling display at the Columbian Exposition that included over 90,000 lamps, a model kitchen with an electric coffee pot and a grill along with a heated saucepan and a chafing dish.

The generators used to power this show were the largest in America. Twelve 1,000 hp, 2 phase, 2,300 volt, 60 cycle generators along with almost a half million feet of wire were very convincing to the public. The dc system seemed to be on its way out. The final push came when, on November 16, 1896, the Niagara Falls Power Company closed the switches and began supplying electricity 22 miles away to Buffalo.

At the turn of the century Tesla was experimenting with high frequency electricity. He established a huge experimental laboratory near Colorado Springs where he built his famous coil. The Tesla Coil is an air core transformer where the discharge of a capacitor through the primary windings develops a discharge in the secondary windings of an extremely high voltage (Fig. 4-5). The one he built here, however, was a gigantic apparatus where the 3-turn primary coil was 75 feet in diameter circling a room. In the center of the room stood his 100-turn, 10-foot-wide secondary. One end of the secondary was connected to earth ground the other end went to a 200-foot mast with a copper ball a yard wide at the top. When the power was turned on the lightning discharging from the ball could be heard for 15 miles.

In 1900 he built a 180 foot tower at Wardenclyffe, Long Island. He believed he could transmit electric power through space around the world. The tower was destroyed at the outset of World War I to keep it from being used by enemy agents.

Tesla died January 7, 1943 without realizing his dream, but his genius led the way into a new world of generating, transmitting, and using electric power. It was an exciting era.

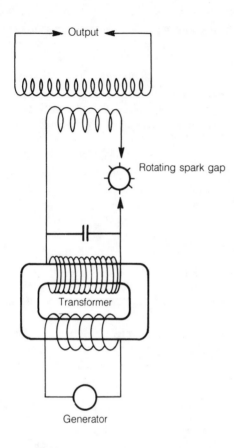

Fig. 4-5. The Tesla Coil. The voltage from the generator becomes stepped up nearly 100 times by the transformer. Sparks created by the rotating spark gap start the circuit oscillating at a very high frequency developing an output of millions of volts, but a very low current.

ELECTRICITY AT THE TURN OF THE CENTURY

Between the late 1890s and 1915 the electric car became popular. Electric cars were typically driven by a single 2 to 5 hp motor powered by lead-acid batteries. These early vehicles were easy to operate, quiet, and could travel up to 20 miles per hour for a couple of hours without recharging. About 35,000 were built in the United States. This same period witnessed elevated electric trains operating in Chicago, Boston, Brooklyn, and New York City.

In 1897 an English physicist, Joseph Thompson (1856-1940), discovered that all atoms contain particles of electricity regardless of the kind of atom. These identical particles he called *corpuscles*. (Today they are known as *electrons*.)

Guglielmo Marconi (1874-1937), an Italian inventor, using similar equipment developed by Heinrich Hertz, was able to send a wireless message about one mile. In 1896 he received a patent for wireless telegraphy. In 1901 Marconi sent the first transatlantic signals from England to St. Johns, Newfoundland, a distance of about 2000 miles.

Soon afterward, John A. Fleming (1849-1945) began conducting experiments with the Edison bulb (Fig. 4-6A) to develop a device to detect these complex

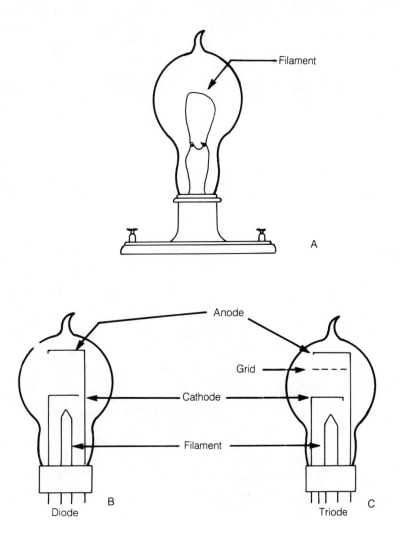

Fig. 4-6. (A) The Edison bulb. (B) Fleming's tube is sometimes called the Fleming valve. (C) DeForest's tube detected and amplified radio waves.

electrostatic and electromagnetic waves we now call radio waves. It turned out to be a diode or rectifier and would convert alternating current into nearly direct current for a more useful purpose (Fig. 4-6B).

An American, Lee DeForest (1871-1961), often thought of as the "Father of Radio" invented a vacuum tube in 1907 that not only could detect the signals but could amplify them to a level where they could be heard through a loudspeaker (Fig. 4-6C). Fleming and DeForest's tubes made radios possible and led to the invention of television in the 1920s, the development of radar by the 1930s and the first electronic computers in the 1940s.

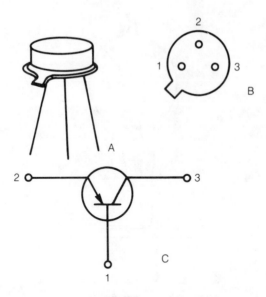

Fig. 4-7. (A) A typical small signal transistor on the market today. (B) Bottom view showing locator tab and pins. (1) Base. (2) Emitter. (3) Collector. (C) Schematic symbol of a PNP transistor.

THE FIRST TRANSISTOR

Two days before Christmas in 1947 three American scientists, John Bardeen, Walter Brattain, and William Shockley, working at the Bell Telephone Laboratories, positioned two tiny wires 2/1,000 of an inch apart on a germanium crystal (a semiconductor). Much to their surprise, this very small device amplified a telephone voice 40 times. This crystal had the ability to perform as a vacuum tube. Bell Laboratories announced the invention of the transistor in 1948 (Fig. 4-7). Shockley, Brattain, and Bardeen received the 1956 Nobel prize in Physics for their amazing discovery.

The efforts of many experimenters over a number of centuries have brought us a long way indeed from the strange behavior of a curious rock called magnetite.

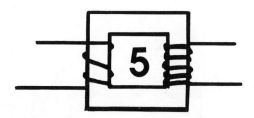

Today's Quiet Giant

Michael Faraday's discovery of the transformer and Nikola Tesla's introduction of alternating current laid the foundation for the electrical services we largely take for granted today. In the past, people had only the power of their muscles to depend on and just surviving was a full time occupation. Before the discovery of fire, humans rarely lived past their teens. Then animals were domesticated, fire was used for heating and cooking, and people were growing some of their food. The average age doubled.

About 200 years or so ago, things began to change. Machines began to take the burden of heavy labor. New medical discoveries were made and the life expectancy again doubled. Today we live much better and work only about one-third as much as our primitive ancestors. This is due solely to the harnessing of electricity. This silent stream of electrons in motion comes to us from huge generators where large magnets are set spinning inside coils of wires. A number of stages are required before you can flip a switch and light a lamp (Fig. 5-1).

We can neither create nor destroy energy. It can only be transformed. It would appear then, that if we convert the energy from coal into electric energy, we would have as much electrical energy as the original chemical energy in the coal, but this is not so. When energy in one form is transformed to another, it tends to move to a less concentrated, more random state. In this manner of conversion, we always end up with less useful energy.

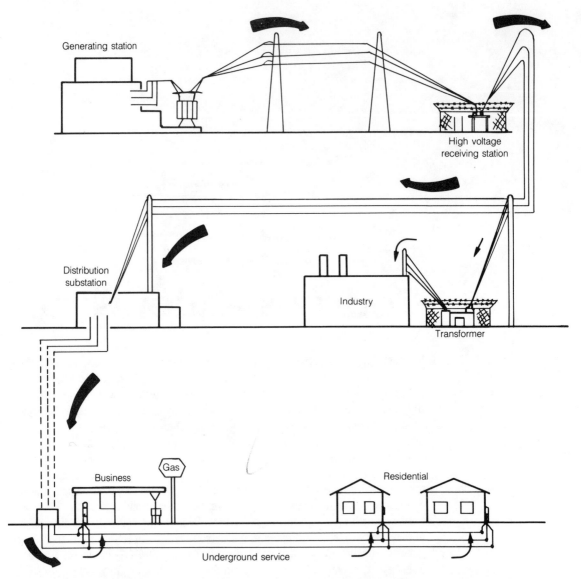

Fig. 5-1. Power distribution today. Electricity leaves the generator at 22-26,000 volts. Passes through a transformer, stepped up as high as 760,000 volts, enters transmission system to receiving station, then is stepped down to about 12,000 volts at substation for further distribution.

ELECTRICITY FROM THE UTILITY COMPANY

Electric utility companies use large turbines to convert the energy of either a liquid or a gas to mechanical energy. This process is performed at the power generating station, often at some site distant from the consumer (Fig.

5-2). Here coal is used to heat water which makes the steam that then turns the blades of the turbine.

Magnets attached to the rotating shaft of the turbine spin inside a stationary ring which is wound with a long piece of wire. As the magnet rotates, the magnetic field induces a small current in each turn of the coil of wire. The magnet sees each turn as an individual electric conductor, but as it is only one, all the small currents are added and the result is a current of considerable size. This current is the product sold to us by the utility company.

In order to deliver the electricity to our homes, the company must maintain a number of different types of facilities. Electricity begins its journey from the generating plant over special wires called *transmission lines* (Fig. 5-3). The nature of electricity dictates that voltage is easier to move than current; consequently these transmission lines carry a very high voltage—and it moves very quickly. Electricity travels at the speed of light.

Fig. 5-2. The Four Corners Power Plant, one of the largest coal-fired plants in the country, generates a major portion of the electricity needed to meet Arizona's energy needs. Other plants may use oil, gas, or nuclear power to generate heat and steam to turn turbines to generate electricity. Four Corners can produce 2,050,000 kilowatts. It is located 15 miles west of Farmington, New Mexico, on the Navajo Indian Reservation. (Courtesy of Arizona Public Service Company.)

Fig. 5-3. This 500 kV transmission line carries electricity directly from the plant's switchyard to a bulk power substation where voltage will be reduced. Voltage must be reduced before it can be used by customers. Lines that carry electricity directly from the plant are usually 345 or 500 kV but there are some transmission lines which carry 765 kV, and experimental lines which carry 1200 kV. (Courtesy of Arizona Public Service Company.)

Fig. 5-4. The Westwing Substation where 500 kilovolts is reduced to 230 and 345 kilovolts. It is jointly owned by Arizona Public Service, Salt River Project, Western Area Power Administration, and Tucson Electric Power. (Courtesy of Arizona Public Service Company.)

Fig. 5-5. Kyrene substation. (Courtesy of the Salt River Project.)

Fig. 5-6. Copper Mine substation. (Courtesy of the Salt River Project.)

These transmission lines continue on from the generating plant to receiving substations (Fig. 5-4). Here transformers like those shown in Figs. 5-5 to 5-8 step down the voltage to be delivered to industrial users and other substations. This part of the distribution system still carries very high voltages as can be seen by the size of the insulators protecting the wires (Fig. 5-9).

The electricity then moves to different areas of the city. It is further stepped down through transformers at more substations. These facilities are usually within the public view, so they are generally built to blend in favorably with their surroundings (Fig. 5-10). These substations often supply local industries with power and depending on the circumstances, these transmission lines may run underground (Fig. 5-11).

As the electric energy moves closer to homes it is once more passed through transformers at residential substations. Again, these distribution stations are usually built with consideration for the surrounding architecture (Fig. 5-12). Once

Fig. 5-7. Orme substation. (Courtesy of the Salt River Project.)

Fig. 5-8. Maryvale substation. (Courtesy of the Salt River Project.)

Fig. 5-9. Often a 230 kV line comes from the bulk power substation, but voltage can also be reduced to 69 kV if needed. Photo shows a 3 circuit 230 kV transmission line (distinguished by the 6 insulators) and a 69 kV line underneath (with 3 smaller insulators). (Courtesy of Arizona Public Service Company.)

Fig. 5-10. Sunnyslope Transmission Substation where 230 kilovolts is reduced to 69 kilovolts and some of the 69 kV is reduced to 12 kV. (Courtesy of Arizona Public Service Company.)

more the electricity leaves a substation and works its way over power lines coming closer to the residential and commercial user (Fig. 5-13).

At this point transformers are again used to lower the voltage to a level that easily supplies most small businesses as well as large residential users (Fig. 5-14). Finally the electrical energy arrives outside our homes to be stepped down by a smaller transformer. Overhead power lines use a pole-mounted transformer such as the one shown in Fig. 5-15 while some residential areas have underground services that use low-profile, pad-mounted transformers as illustrated in Fig. 5-16.

Mounted above the service's entrance to the residence, the utility company installs a meter to measure the amount of electricity consumed in the home. These are called watt-hours or kilowatt-hours meters, and these too depend on the transformer principle of induction to measure the power used (Fig. 5-17). Basically this ac watt-hour meter is an aluminum disc mounted horizontally in between two electromagnets. The disc is turned by the induced currents of the two magnetic fields. The strength of the field from the top electromagnet is proportional to the supply voltage while the strength of the lower field is relative to the load current. The force that turns the disc is then proportional to the product of the voltage and current. As this is the driving force, the rotating speed

Fig. 5-11. The 69 kV line can supply electricity to a large user like an industrial plant. To be used by a typical residential user, voltage must again be reduced via a distribution substation. Underground lines are more expensive to install and repair than overhead ones, but for aesthetic reasons or lack of overhead space, lines are sometimes put underground. (Courtesy of Arizona Public Service Company.)

Fig. 5-12. Mountain View Distribution Substation where 69 kilovolts is reduced to 12 kilovolts which then travels on lines through residential areas. The low profile substation is designed to fit in with the neighborhood. (Courtesy of Arizona Public Service Company.)

of the disc is governed by the amount of power consumed by the load. The accuracy of these meters is typically within about 2 percent. Usually at monthly intervals, the utility company sends out meter readers to record each consumer's power usage. This determines your monthly electric bill.

BACK-UP SYSTEMS

As you flip on lights and turn on appliances, the generating plant instantly sees this increase in electrical consumption and must respond with a larger supply. Unlike water stored in a reservoir or behind a dam, there is no way to store electricity in large amounts. As a consequence, the suppliers of our electricity must constantly maintain a small army of personnel and equipment to generate electricity the instant the demands are made. Since power is generated on demand, if the generating plant suddenly ran out of fuel, the entire operation would come to a halt. It immediately becomes obvious why the utility companies are so concerned with the problems of fuel supplies and their costs.

Most generating plants keep some form of generating ability on stand-by; one generator turning but without producing power. This provides some insurance that, if needed, this reserve can be brought into action. Also, there

Fig. 5-13. The 12 kV line is often the last distribution line before voltage is reduced by a transformer and used by a residential customer. Most lines going through residential neighborhoods are either 12 kV or 7.2 kV lines. (Courtesy of Arizona Public Service Company.)

Fig. 5-14. This three phase overhead transformer can transform 12 kV or 7.2 kV into 120, 240, or 480 volts—enough power for a larger residential customer or a typical commercial customer. (Courtesy of Arizona Public Service Company.)

Fig. 5-15. This overhead transformer on a 7.2 kV line converts power to 120 or 240 volts, a normal current for household use. (Courtesy of Arizona Public Service Company.)

Fig. 5-16. This pad mount transformer performs the same function as an overhead transformer but is connected to underground cable. A transformer usually serves 4 or 5 homes. (Courtesy of Arizona Public Service Company.)

Fig. 5-17. Electric meter.

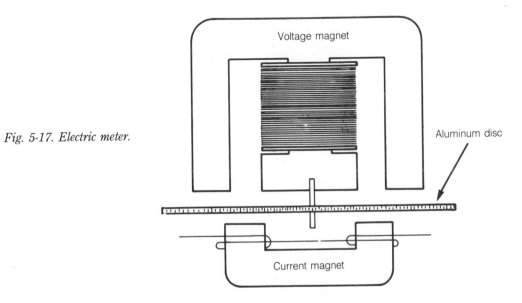

Voltage magnet

Aluminum disc

Current magnet

exists a cooperative system between utility companies in America that comes into play in the event one area has a blackout from lack of fuel or because of some other problem. For example, there is a Western region network of interconnected systems, creating a grid of power across our nation. This means that if one utility company developed a problem generating enough electricity, the other companies would step in and furnish the necessary power to the disabled utility until the problem was cured.

Every day electricity plays a major role in our way of life and every day our electrical needs continue to grow. Electricity remains one of the most useful, adaptable, and clean energy sources available.

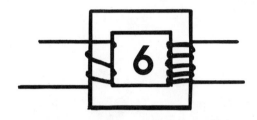

Magnetism
in Industry

James Watt started the industrial revolution with his steam engine and relieved man and animals from the burden of much heavy labor. Today, through the wonders of electromagnetism, electric motors driving hydraulic and pneumatic circuits can do both heavy and light precision work. For example, magnetic cranes can be seen operating in scrap yards and salvaging iron from city dumps.

ELECTROMAGNETIC DEVICES

The magnetic induction of transformers delivers a workable voltage to manufacturing plants that drives a variety of electric motors from huge 100 hp air-handling motors that circulate air for employee and equipment comfort, to the small motors in automatic pencil sharpeners on a drafting desk. The larger motors are even turned on and off by magnetically controlled relays in motor control centers. Basically, a relay is a switching device that mechanically opens or closes contacts when its coil is electrically energized or de-energized. The simplest of relays operates by electromagnetic attraction. This relay consists of a coil, plunger, and a set of electrical contacts (Fig. 6-1). When the current flows through the coil, a magnetic field is developed that causes the plunger to quickly move and close the contacts. Motor starters are relays that turn the motor's power circuit on and off. Overload relays disconnect the motor-starter coil if some mishap causes the current to get too high. Control relays can be found inside most equipment's electrical panels performing switching jobs that allow the equipment to operate.

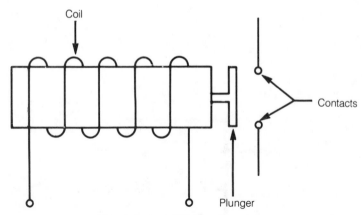

Fig. 6-1. A simple plunger relay.

There are also *time delay* and *latching* relays. The time delay relay can open or close its contacts at a predetermined time interval after it has been actuated or deactuated. The latching relay uses its input signal to latch, closing or opening the contacts. It stays in this mode even after the signal is removed. This relay requires a second signal to unlatch.

Another hard working electromagnetic device found throughout any industry is the solenoid. The solenoid is an electric coil containing a plunger. When the coil is energized, the plunger moves into the coil allowing it to do mechanical work. When the coil is de-energized, a spring or gravity returns the plunger to its normal position. These handy devices are also found in valves controlling the flow in hydraulic and pneumatic circuits.

Even a few fluids can be moved by magnetic fields. Some nuclear reactors use a liquid metal as a heat-transfer media. Electromagnetic pumps have the advantage over conventional mechanical pumps because they have no moving parts, bearings, or seals, to wear out, almost eliminating maintenance. They operate on the principle that a certain amount of force is exerted on a current-carrying conductor in the presence of a magnetic field. The liquid metal is the current-carrying conductor. There are two types of pumps: direct-current conduction pumps and alternating-current induction pumps. The dc pumps are a good example of the right-hand rule which states that any current moving at right angles to a magnetic field develops a force at right angles to both.

Any current in a conductor has a magnetic field. If this conductor is placed in another magnetic field, the combined fields react and produce a force or motor action (Fig. 6-2). Pump performance depends on the amount of current and the strength of the magnetic field. These pumps have been fully developed and large commercial units are in use.

PERMANENT MAGNETS IN USE

Permanent magnets are manufactured for use in almost any industry from

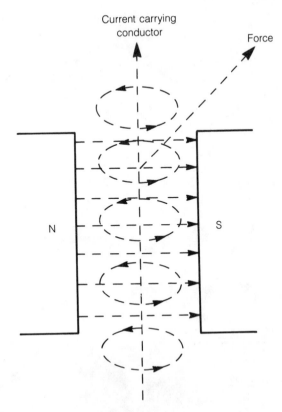

Fig. 6-2. Force or motor action in a current-carrying conductor inside a magnetic field.

ore separation to food processing. Machinery such as grinders, hammermills, shredders, augers, and impellers can be protected from damage by positioning a magnet so it removes large tramp iron such as bolts, chain and even accidentally dropped hand tools.

A grain producer might use something like a plate magnet (Fig. 6-3A) as insurance against metal contaminants in the product.

Material traveling on a conveyor belt can be magnetically cleaned by installing a magnetic pulley (Fig. 6-3B) or a drum magnet (Fig. 6-4A).

Selective high gradient magnetic separators are in use that can separate metals of different magnetic properties.

Physicists have known for a long time of the vast number of elements and their compounds or minerals that are paramagnetic. However, since the 19th century only strongly magnetic materials such as iron or magnetite could be profitably separated from nonmagnetic materials.

In a process where paramagnetic materials are the contaminants, expensive chemical leaching is used to remove the particles. Mining "china clay" or kaolin involves such a process. Kaolin is a pure white clay made of decomposed feldspar.

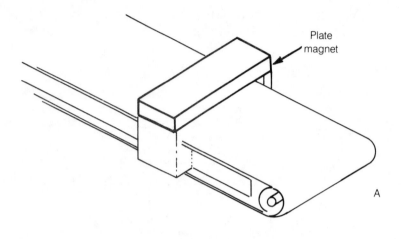

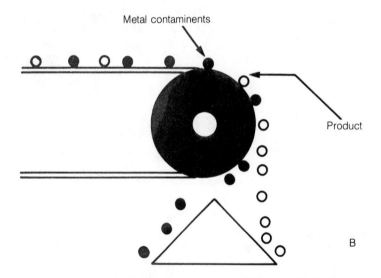

Fig. 6-3. (A)Plate magnet mounted over a conveying belt will pull metal up and out of the product. (B) A magnetic pulley not only removes metal contaminants but is self-cleaning, which eliminates maintenance.

When ground into a fine powder, it is widely used for making the finest pottery. In 1968 the first paramagnetic separator was built and available for commercial use. Slurries of kaolin are exposed to a high intensity magnetic field for controlled retention times of 30 seconds or more, effectively removing the paramagnetic impurities.

Large, 84-inch paramagnetic separators have been capable of processing over 100 tons of kaolin per hour for the last decade. The efficiency of this separator might have doubled the United States reserves of kaolin. Now, every major kaolin producer in the United States and England has at least one paramagnetic separator—some even have two or more. Paramagnetic separators revolutionized the processing of kaolin for the mining industry.

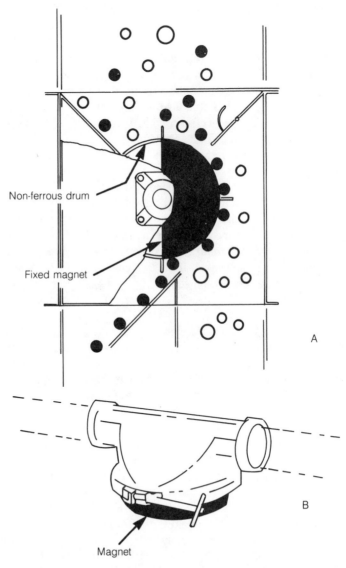

Non-ferrous drum

Fixed magnet

A

Magnet

B

Fig. 6-4. (A) The drum magnet consist of a non-ferrous drum rotating around a fixed magnet. (B) The magnetic trap is an effective way for removing stray metal from a fluid processing line.

Beverages, crushed fruits and vegetables, meat spreads, and ground meat products are often protected by a USDA-approved magnetic trap installed in the line just prior to packaging (Fig. 6-4B).

Magnetic sweepers eliminate back-breaking clean-up for roofing and other contractors who otherwise might leave hazardous nails and scrap metal on floors and yards. A magnet attached to any yard vehicle can save equipment down time and money spent on tire repairs.

METAL DETECTORS

Metals have one property that distinguishes them from all other elements and allows them to be detected by a sensitive device appropriately called a metal detector. This property is the metal's ability to conduct electricity. When a varying magnetic field is generated near a piece of metal, an electric current is induced in the metal. This induced current in turn establishes its own magnetic field around the piece of metal. This second magnetic field tends to disturb the first magnetic field. A metal detector is simply a device that can see this disturbed magnetic field.

Metal detectors that operate from an induced magnetic field include balanced search coil units, field search units, and pulsed magnetic units.

Balanced coil search units are made with two coils that have the same number of turns. Each coil has a primary and a secondary winding with the primaries operating in series. When an alternating current is applied to the primary, a varying magnetic field is developed. If the two coils are in an area without any metal, the induced field seen by the secondary windings is the same and as the secondary coils are connected in such a manner that they oppose, there is no signal.

If a metal object is in the area, it develops an induced magnetic field which is felt by the nearest secondary coil. This signal is compared with the other coil signal and the difference is the signal that indicates a metal object.

Metal detectors that employ the principle of magnetic induction are limited to a range of about six inches when used over the ground because differences in soil conductivity can easily alter the detector's readings.

The heterodyne search unit is another metal detector that uses the principle of magnetic induction. It also uses two coils but each has only one winding. An oscillator is part of each coil's circuit. Initially, each oscillator is set to the same frequency so that both coils produce a varying magnetic field at a specific frequency.

If one of the coils is placed near a metal object, the metal affects the inductance of that coil which in turn alters the frequency of oscillations in that coil's circuit. Now the two oscillator circuits compare signals and as one is off, a beat frequency is produced. This is the signal that indicates the presence of metal. This detector can also be affected by the soil's conductivity.

Field search detectors operate much the same as balanced coil detectors except that they use a wire loop operated by a high powered oscillator. This greatly increases the range because it also senses the soil's conductivity. With this type of detector, archaeologists are often able to locate the outlines of ancient walls and buildings.

Pulse magnetic detectors operate on the principle that a magnetic field travels at a measurable speed: the speed of light. These units have a search coil that bounces a signal off metal objects. The search coil transmits a brief, high-powered pulse, and then switches to a receive mode. When the pulse strikes a metal object, it induces a magnetic field in the metal which is detected and compared with the time of the transmitted pulse.

Because ferromagnetic materials have such a high permeability, they can be detected by a different method. Earth's magnetic field has lines of flux which tend to travel the paths of least resistance. These flux lines, therefore, converge in areas of any ferromagnetic material causing the overall magnetic field to be distorted. Everything on Earth is within the magnetic lines of the Earth's field. A device called a magnetometer, can detect this distortion. An airborne unit was developed in World War II to detect the distorted magnetic field created around submarines.

MAGNETISM USED IN PRODUCTION

Manufacturers that need to coat their product with a thin layer of metal often use a magnetically-enhanced method of metal deposition called *sputtering*. The product can be of almost any material including glass, other metals, and many types of plastic. Some optical coatings for glass and even friction-reducing coatings for razor blades benefit from this process. That shiny chrome ornament on your car may be plastic underneath.

In the electronics industry, the production of integrated circuit chips depends heavily on this process. The chips are not made individually but by the batch, sometimes hundreds on a single silicon wafer four to six inches in diameter and about 15 mils thick. The wafer is first polished to a mirror finish, and then a thin layer of glass is applied for an insulator.

Figure 6-5 shows a piece of equipment called a *diffusion furnace*. These furnaces operate at about 1,200 degrees centigrade and are typical of the machines used to grow some of the layers on the wafers. The patterns for the individual circuits are reduced and reprinted in rows over a small glass plate called a *mask*. The mask is then used to expose the pattern on wafers coated with a photosensitive layer (Fig. 6-6).

Next, unwanted portions of the pattern are etched away by acids (Fig. 6-7). In these production areas where acids are used, sealed magnetic switches can be installed in tanks as safety interlocks where the acid would make short work of a conventional switch (Fig. 6-8). Other magnetic lid, or cover, interlocks are used as a safety precaution to keep equipment from operating if the covers are not locked or in place.

Certain elements are used to establish a desirable electrical characteristic

Fig. 6-5. Diffusion furnace. (Courtesy of Motorola Semiconductor Products, Inc.)

in the silicon wafer. Machines called ion implanters were designed to implant these elements in a very precise manner (Fig. 6-9).

Ion implanters use a series of magnetic fields to guide a high current beam, under vacuum, to the wafer to be implanted (Fig. 6-10). The ion source is gas-fed. Here the ion beam is extracted from the source and accelerated to the desired implant energy level. Next, the beam is bent through a 60 degree arc by an analyzing magnet. The desired ions follow the curved path and the unwanted ion species follow a different trajectory and strike graphite surfaces called beam dumps. The analyzer magnet also focuses the beam in the vertical plane.

The beam next passes through a second magnetic field where it is focused in the horizontal direction then magnetically deflected into the implant chamber. This second deflection also helps clean up the beam by removing unwanted particles. Inside the implant chamber, called the *end station*, the desired ions are implanted into the wafers.

Next, depending on the type of circuit, a layer of metal such as aluminum,

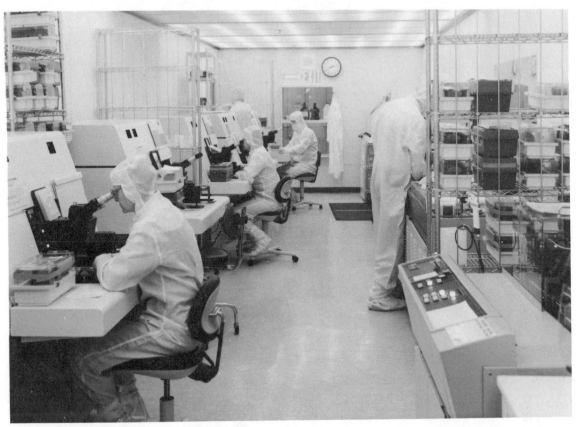

Fig. 6-6. Operators aligning and exposing wafers. (Courtesy of Motorola Semiconductor Products, Inc.)

titanium, platinum, or tungsten is deposited. The process is repeated with insulators and conductors in thin films until the desired circuits are achieved. Machines such as the one shown in Fig. 6-11 are used for this purpose.

Sputtering is performed inside a vacuum chamber where a trace of an inert gas, usually argon, is present (Fig. 6-12). The source of the coating material is called the *target* and is connected to the negative terminal of a high voltage supply and becomes a cathode.

The wafer surface is called the *substrate* and is the base for building the transistor or IC (integrated circuit). This process uses a plasma glow discharge, a phenomenon similar to that found in a neon sign or a fluorescent light. When the power is turned on, electrons streaking from the cathode strike the argon atoms, knocking away some electrons and creating positively charged ions. These positive ions are then sped toward the target by the power supply voltage. They hit the target with enough force that they knock loose small molecule-size particles of the target material.

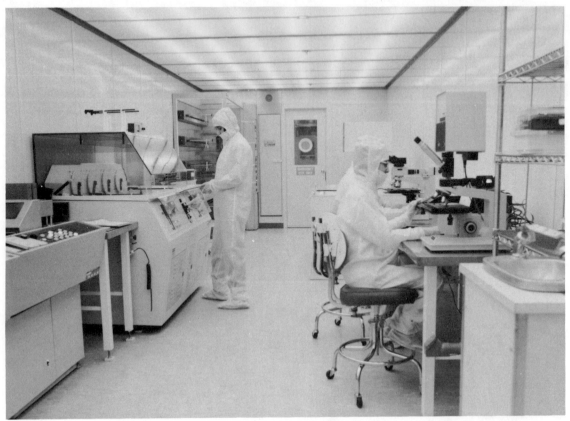

Fig. 6-7. Operator standing at an etch station. (Courtesy of Motorola Semiconductor Products, Inc.)

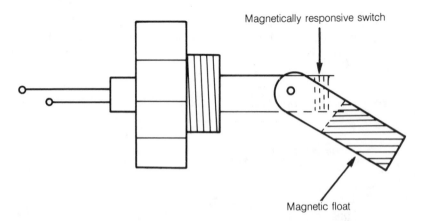

Fig. 6-8. A liquid level switch can be used to disconnect a circuit, such as a heater, when the tank is empty.

Fig. 6-9. Ion implanter. (Courtesy of Motorola Semiconductor Products, Inc.)

These small particles leave the target at nearly the energy levels of the striking ions and then go on to strike the surface of the substrate with a force strong enough to form a thin coating with a very high adhesion characteristic. By locating magnets around the target in a special arrangement, the magnetic field contains the electrons in a dense cloud—a magnetic bottle. This enables the trapped electrons to strike more of the gas molecules and increase the ion density in the plasma which increases the number of ions bombarding the target and raises the rate of deposition.

The glow discharge in this process is an effect very much like the northern lights, in that it is caused by incomplete ionization of all the gas molecules in the electric field. When electrons collide with gas molecules, some of the electrons orbiting the gas molecule are not knocked completely free. Instead they absorb energy and are raised into a higher orbit. But this orbit is unstable, and when the electron moves back into its stable orbit, it releases the energy it took on as a photon of light. The color of this glow depends on the kind of atoms that make up the gas molecule and the amount of power involved.

Most vacuum chambers have quartz windows, or ports, that allow the operator to monitor the process. These ports are normally covered with

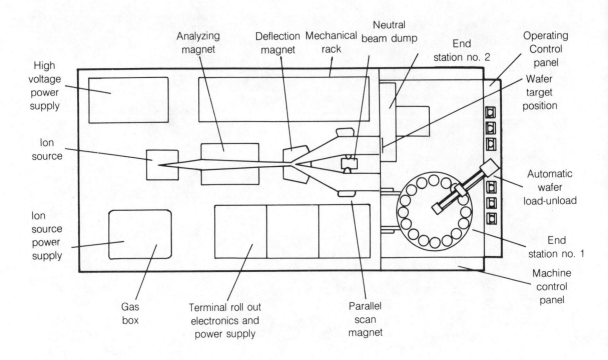

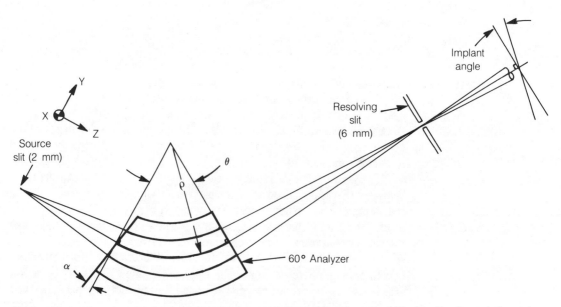

Fig. 6-10. The Varian model 160-10 ion implanter. (Drawings courtesy of Varian Equipment Group.)

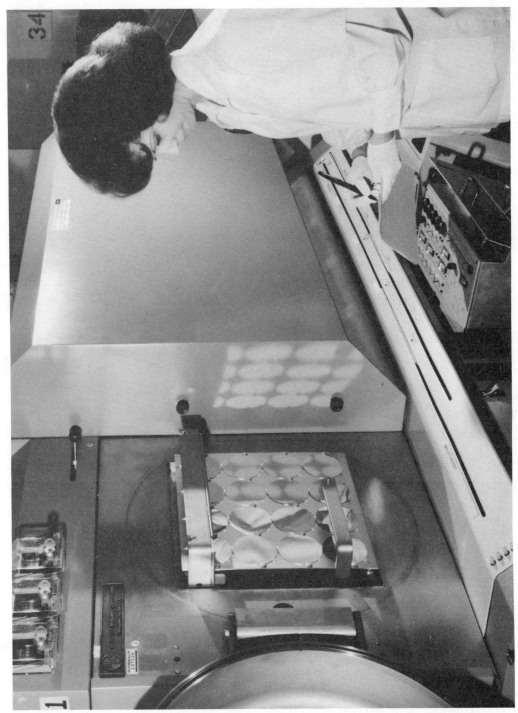

Fig. 6-11. Completed wafers on pallet ready to be removed from vacuum load-lock. (Courtesy of Motorola Semiconductor Products, Inc.)

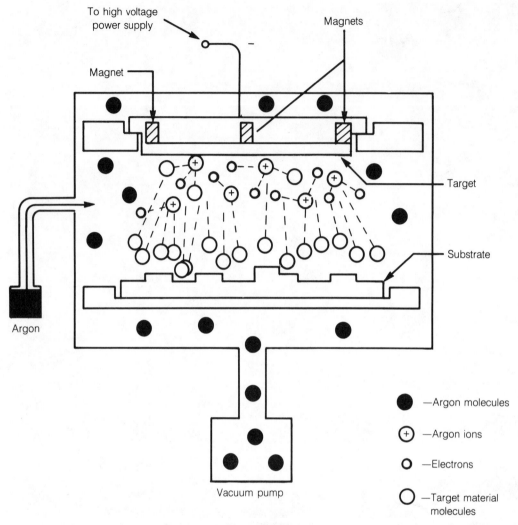

Fig. 6-12. Thin film sputtering.

protective shields inside the chamber. To open or close the shield some kind of vacuum tight feed-through must be furnished. Ferrofluid liquid seals are sometimes used. A ferrofluid is a gelatinous, glue-like suspension of tiny magnetic particles in a liquid carrier. These particles are coated with an agent that prevents them from balling up even when a strong magnetic field is applied. Outside a magnetic field, the magnetic moments of the particles are haphazardly distributed and the fluid is not magnetized. When a uniform magnetic field is applied, the particles experience only torque and almost instantly align themselves with the field lines. But in a gradient magnetic field, the particles feel a force that causes them to respond as a uniform magnetic liquid and move to the area of the strongest field. This allows ferrofluids to be accurately positioned and controlled by an outside magnetic field.

Ferrofluids are classified as a soft, superparamagnetic material and are used in many industrial and consumer applications. They are used in inertia dampers to increase the performance of stepper motors. These dampers reduce the settling time of the shaft and almost eliminate torsional vibrations. Ferrofluids are often used in the air gap around the voice coil of loudspeakers as well as in the inspection of the critical dimensions on turbine blades. In the sealing device, the rotary shaft seal employs a magnetic liquid barrier which produces an excellent vacuum seal with very low torque drag. A permanent magnet and a magnetically permeable focusing arrangement produces a completed magnetic circuit around the rotating shaft. These ferrofluid feed-throughs are devices that can transfer rotary power from normal room conditions into a controlled location such as the vacuum chamber.

OTHER USES OF MAGNETISM

Magnetic treatment of water has been used for decreasing scale and salt formations in steam boilers, pipelines to heat exchangers and other industrial circulating systems. This treatment consists of using one or more pairs of permanent magnets or electromagnets mounted on the pipe in such a way that the water flows between the poles of the magnet. A magnetic water filter.

The magnet has even found its way into surgery. A metal splinter in an eye could mean a very delicate operation, but hospitals are often able to use an electromagnet to extract the splinter. This procedure lessens the hazard of the operation as well as the cost and discomfort for the patient.

Near Miyazaki, about 570 miles southwest of Tokyo, the Japanese National Railways have been experimenting with a train that floats on a magnetic field at 300 miles per hour. By December 1979 this amazing 44-foot long car had attained a speed of 323 MPH. The power source comes from 16 huge electromagnets. Half are used for levitation and the other half for propulsion. Four 14-foot long L-shaped chambers contain the magnets. Cryogenic temperatures tend to remove the resistance in electrical conductors so technicians pump liquid helium into each chamber lowering the temperature of the magnets to about minus 260 degrees celsius.

The eight propulsion magnets are mounted vertically and the eight lifting magnets are laid horizontally, corresponding to a similar arrangement embedded in the tracks. An automatic signal turns on an electromagnet in the vertical section of the track in front of the car. The vertical magnets in the chambers react and pull the car forward. The magnetic field in the track then advances. The sequence is repeated as the on-board magnets again pull ahead to catch up. The magnetic field in the track leaps from coil to coil causing the vehicle to accelerate. The vehicle does have wheels, but they are useful only in starting up or in braking. When the car begins to move, the horizontal magnets go to work. As they pass

Fig. 6-13. The Magnequench magnet is available in a variety of strengths, depending upon its application. Compared here with a U.S. quarter, the magnet used in Delco Remy's new permanent magnet gear reduction cranking motor (starter) measures approximately 1¼" × ¾" × ¼", and weighs slightly more than ¾ oz. The motor uses six of these new magnets. (Courtesy Delco Remy)

over the horizontal fields in the track, the induced current creates a magnetic field of the same polarity as the ones in the car. Like poles repel. The greater the speed, the stronger the magnetic field until the car is floating on a magnetic cushion 4 to 5 inches high. These new trains could significantly reduce noise

Fig. 6-14. Magnequench is credited with reducing the size and weight of Delco Remy's current cranking motor (starter) for passenger car and light-duty truck applications by half. At left is the new permanent magnet gear reduction cranking motor (PMGR) introduced in the 1986 model year, compared with the 10-MT cranking motor it replaces (on the right). (Courtesy Delco Remy)

levels as well as pollution, but we will have to wait to see what effect these strong magnetic fields have on the passengers and crews.

In the 1890's, Frank and Perry Remy started a home wiring business in Anderson, Indiana, but there were almost a dozen automobile manufacturers in town that offered better opportunities. By 1910 they had developed a small ac generator used to supply the ignition system and were producing 50,000 of these magnetos a year. One of their competitors, Dayton Engineering Laboratories Company (Delco), was also making ignition equipment and generators for auto and marine use.

In 1918 these two companies joined the General Motors Company and shortly after World War I were manufacturing batteries under the Delco Remy banner. In March 1985 plans were announced to construct a 160,000 square foot facility to manufacture a new magnet. The "Magnaquench" process was discovered by General Motors scientists at their research laboratories in Warren, Michigan. Magnaquench is a new high performance magnet manufactured from neodymium, iron and boron.

This new magnetic alloy allowed engineers to replace the traditional copper wire coil with smaller and lighter permanent magnets (Fig. 6-13). This means the new starters are about half the size and weight of the older ones. The new magnet is used in GM's permanent magnet gear reduction cranking motor (PMRG) (Fig. 6-14). This new starting motor was introduced in a few of the 1986 GM cars and light-duty trucks; however, this new magnet also has potential for many uses in other electrical and magnetic devices.

It would not be an exaggeration to say that all of the electricity used in industry is based on the generation and utilization of magnetic fields.

Industry world-wide is so heavily dependent on magnetism and electromagnetism that without this phenomenon, we would have a technology that peaked with signal fires for communication and whale oil lamps for lighting. Fortunately, physicists continue to discover new uses for this mysterious force and certainly many of these new applications will go to improving the efficiency of machines as well as the safety of employees.

Dc Circuits

Although electric utility companies provide us with a good supply of alternating current, a large number of motors, welders, and electronic devices use direct current in their operations.

ELECTROMOTIVE FORCE

Electric current, the flow of electrons in a circuit, is measured in units called amperes. For current to flow it must have a complete path or circuit (Fig. 7-1). There also must be some kind of force or pressure to push these electrons along. This electromotive force (emf) can be developed from batteries, generators and other devices. This electrical pressure or emf is measured in volts and, like a higher water pressure, causes more gallons per hour to flow in a pipe, a higher voltage causes a greater current to flow in a wire.

With any flow there is some resistance and with any flow the best conductors have the least resistance. Copper is the most common conductor for electron flow, but to be a useful circuit it has some kind of a load which appears in the form of resistance.

This resistance is measured in ohms. One ohm of resistance in a circuit supplied by one volt allows one amp of current to flow (Fig. 7-2).

The electron flow illustrated in Fig. 7-2 is direct current because the current flows in just one direction. The reason for this is that the battery keeps the same polarity. The dc voltage source can vary its output voltage, but as long as it maintains its polarity, electrons flow in one direction and there is a direct

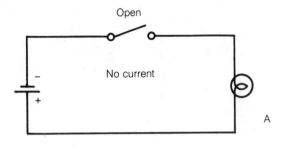

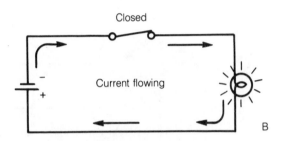

Fig. 7-1. Current requires a completed circuit to flow. (A) Open circuit; no current. (B) Closed circuit has current flow.

current. In alternating current the supply voltage periodically reverses in polarity (Fig. 7-3). The voltage polarity and current direction in homes have 60 cycles of reversal each second.

Batteries do provide us with direct current in a vast number of applications, but they tend to be costly and need replacing or recharging. In most cases alternating current is the only way to have electrical energy supplied at a reasonable cost, but dc is still needed; for example, in vacuum tubes or transistors in radios, TVs, and audio amplifiers.

CONVERTING AC TO DC

When direct current is required you can easily change alternating current to direct current with a handy device called the rectifier. These can be in the form of semiconductors or electronic tubes. A semiconductor can be thought of as a material midway between a good conductor and a good insulator.

Silicon and germanium are semiconductors. The atoms of these elements have four electrons in the outer shell, but an interesting thing happens when

Fig. 7-2. Georg Ohm found that the amount of current depends on the amount of supply voltage divided by the resistance in the circuit.

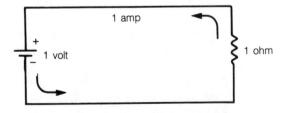

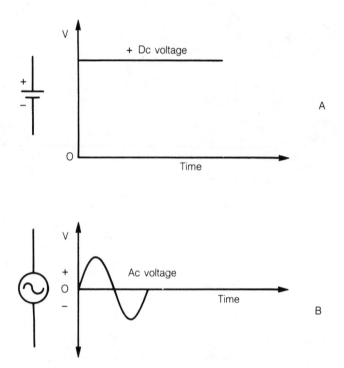

Fig. 7-3. Comparing a dc voltage with an ac voltage. (A) A dc waveform showing a steady voltage with one polarity. (B) An ac waveform or sine-wave, showing an alternating voltage reversing polarity.

we add a tiny impurity. If we mix in an element, such as arsenic, whose atoms have five electrons in the outer shell, four of the electrons fit but one is left to ramble around. This mixture is called an *N-type semiconductor* (N refers to the negative free electrons).

However, if you add an element such as boron whose atoms have only three electrons in the outer shell, it is short one electron and leaves a hole. This is called *P-type semiconductor*. It has no free electrons.

If you join a piece of N and P material, you'll find that a high resistance or insulating barrier has developed at the junction. When a battery is connected one way, current does not flow (Fig. 7-4A), but if it is connected with the other polarity (Fig. 7-4B), current does flow. The first connection has *reverse biasing*, while the second connection has a *forward bias*.

A diode placed in a circuit creates dc current in the form of pulses because it only conducts half of the ac cycle (Fig. 7-5). This might work fine for something like a battery charger but would create a terrible buzz in an audio amplifier. To smooth this dc current a bit, a capacitor can be connected across the load lines (Fig. 7-6), but normally a filter network called a *full wave rectifier* is used to supply a steady dc current (Fig. 7-7).

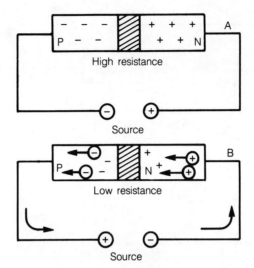

Fig. 7-4. P and N junctions of a semiconductor. (A) Reverse bias with no current flow. (B) Forward bias with current flow.

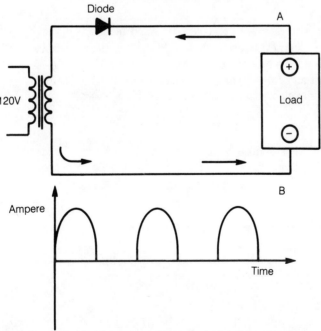

Fig. 7-5. A pulsating dc current in a half-wave rectifier. (A) Circuit with 120 volt step-down transformer. (B) Dc current plotted against time.

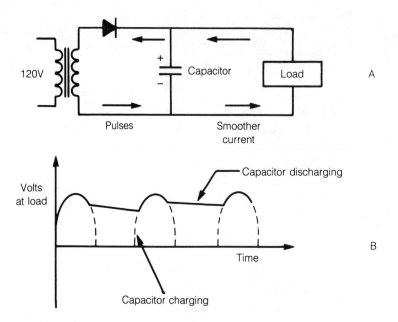

Fig. 7-6. A capacitor smooths the output voltage—somewhat. (A) Improved half-wave rectifier circuit. (B) Graph of smoother dc voltage felt by the load.

Some semiconductor materials are *photoelectric* which means that they convert light energy into electrical energy when light strikes their surface. *Photovoltaic* or *solar* cells do this by producing a voltage and use silicon doped with arsenic for the N part and a thin layer of boron added for the P end (Fig. 7-8). Solar cells typically develop about a half volt.

INDUCTANCE

Dc generators are another source of direct current. Basically, the dc generator consists of rotating coils inside a stationary magnetic field. A copper wire moved quickly across the lines of a magnetic field produces a voltage across

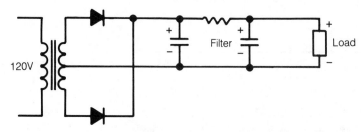

Fig. 7-7. One form of a full-wave rectifier with a center-tapped secondary on the transformer.

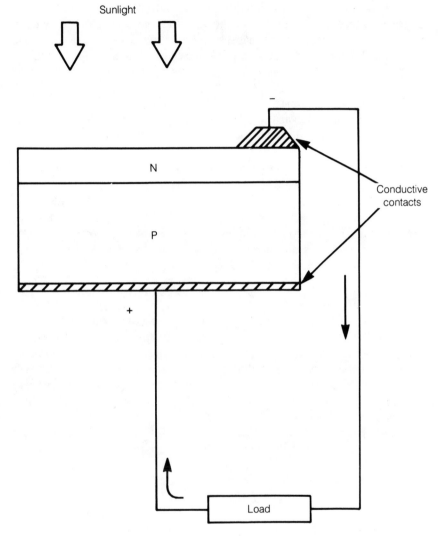

Fig. 7-8. Typical solar cell construction
converts light energy directly into electrical energy.

the ends of the wire. This voltage, or electromotive force, tries to move the electrons of the copper atoms in the wire, but no current flows unless the wire is part of a completed circuit.

The same emf is developed by abruptly sweeping the magnetic field across stationary wires. A stationary coil produces an emf when a bar magnet is quickly inserted or withdrawn from inside the coil. The induced voltage is proportional to the strength of the magnetic field, the speed of the magnet, and the number of coils in the wire. It can be produced by the physical movement of either the magnetic field or the conductor. However, if the current in a conductor is increased, a magnetic field expands outward from the conductor and when it decreases, the magnetic field collapses back into the conductor. As this field expands and collapses corresponding to the changes in the current, it creates the same effect as movement of the conductor and produces a self-induced voltage.

The current does not have to reverse directions; this occurs when a dc circuit is simply turned on and off. This induced voltage is a result of the change in the current and not the value. A steady, unchanging direct current is not affected by inductance.

For calculations, inductance uses the symbol L referring to the linkages of the magnetic flux and a unit of inductance is called the henry. The induced voltage divided by the amount of current change in amps per second will be the number of henrys, or inductance value of the coil (Fig. 7-9). An rf coil, called a *choke*, may have an air core or an iron core. Values for air-core chokes are in millihenrys (mH) and microhenrys (μH) while iron-core chokes may have a value of about 1 to 25 henrys (25 H).

ELECTROMAGNETS

Probably one of the first applications of electromagnetism was the electromagnet. Electromagnets generally have a soft-iron core that becomes strongly magnetized when a current passes through a coil or wire wound on the core. Hard materials are used for permanent magnets while soft materials allow the magnetic field to be turned on and off. Electomagnetism finds its way into many applications such as solenoids, relay coils, electromagnetic brakes and clutches as well as huge lifting magnets.

Joseph Henry gave the world its first powerful lifting magnet. The principle remains the same today, although they are a bit more refined. They are usually constructed with the coil almost completely surrounded by iron, with one pole of the magnet being the core inside the coil and the other pole formed on the shell surrounding the coil (Fig. 7-10). Nonmagnetic manganese steel is often used as a protective cover plate or bumper. Currents in a lifting magnet tend to be large, typically 10 to 20 amps, and the circuit is highly inductive so control can be a problem. If the supply line switch were opened a destructive arc would occur because of the phenomena called *counter emf*. However, the controller

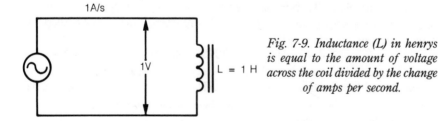

Fig. 7-9. Inductance (L) in henrys is equal to the amount of voltage across the coil divided by the change of amps per second.

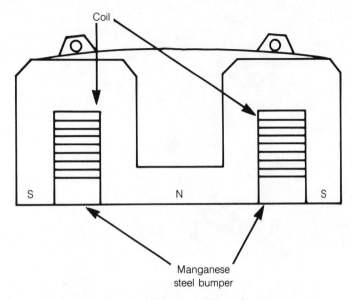

Fig. 7-10. Cross section of a lifting electromagnet.

used to operate a lifting magnet normally takes care of the problem by reducing the current after initial lift to keep the magnet from overheating and introduces a discharge resistor across the magnet before the operator is allowed to turn the switch off.

This automatic controller also causes a low current of reverse polarity to flow in the coil briefly after the switch has been opened. This cancels the residual magnetism and releases the scraps and small chunks that otherwise might have continued clinging to the magnet.

RELAYS

The relay is another form of electomagnetic operation. It is a switch, usually with multiple contacts, that can be closed or opened when current flows through the coil of the relay (Fig. 7-11). When a small current flows in the coil, the relay is activated and the magnetic field attracts the iron armature which is connected to the movable switch contacts. These contacts are under spring tension, pressing against the top row of fixed contacts. When the coil is activated, the armature pivots and the spring-loaded contacts move to connect with the lower row of fixed contacts. The next lower row of fixed contacts can be connected to the circuit under control. In this case, four separate circuits can be turned on or off. The bottom two fixed contacts are for the supply voltage to the coil. The threaded screw between the coil contacts prevent the relay from being plugged in upside down. The coil construction normally will allow for dc or ac operation with coil voltages of 12V, 24V and 120V.

A

Fig. 7-11. A typical relay used in today's circuits. (A) Top view. (B) Bottom view. (Photos courtesy of Jess Castellano.)

B

The relay is considered a separate switching device, and the current and voltage operating the relay does not enter the controlled circuit. Relays are used in many control applications where the small current needed to activate the relay is provided by a device that senses a change in light or temperature. The blower motor in a home's heating and cooling system is a good example. Relays can also be found controlling switches in distant locations or in locations considered inaccessible because of lack of space or a danger from high-voltages or temperatures.

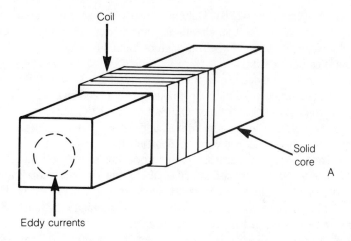

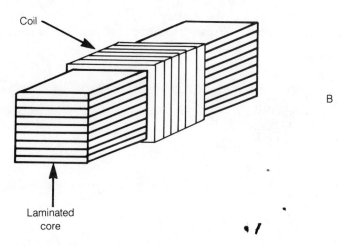

Fig. 7-12. Comparing solid and laminated iron cores. (A) Solid core with troublesome eddy currents. (B) Laminated core without eddy currents.

MAGNETIC FIELDS

A number of dc electromagnets use cores of solid bars or rods. However, for magnetic fields in motion such as dc motors and generators along with ac equipment, the cores are made up of thin sheets of iron. When a magnetic field passes through a wire or any metal, it tends to generate a current in the metal (Fig. 7-12A). These eddy currents are caused by the changing current or the moving magnetic field. Eddy currents are unwanted because they take energy from the coil and create heat in the core. Eddy currents can be greatly reduced by using a laminated core where the laminations are lacquer coated to minimize electrical contact (Fig. 7-12B). Only small eddy currents can circulate within these laminations. These thin sheets are normally made from iron containing a small amount of silicon. This iron mixture has high electrical resistance but still has high permeability.

In physics, and indeed in all things, a general law is that accomplishment of work is in direct proportion to the amount of effort and inversely proportional to the obstructions or restrictions. Ohm's Law, $I = E/R$, states the same thing when it says that the current developed is in direct proportion of the electromotive force and in inverse proportion to the resistance.

You can apply this general law in the production of magnetic lines of force, or flux. The amount of applied effort would be the amount of magnetomotive force developed by the coil. The resistance of magnetism is called reluctance, the opposite of permeability. Air, for example, has a high reluctance while iron has a low reluctance and is highly permeable.

Any magnetic device has a magnetic circuit. A magnetic circuit can be thought of as the path of the lines of force through the device. The circuit can be traced by starting at any point and following the lines of force through the iron then the air and arriving back at the starting point. Due to the high reluctance of air, the magnetic circuit's air gaps should be kept as close as possible.

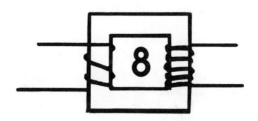

Ac Circuits

Alternating current is the most common source of electric energy because of the unique ability of the voltage to be stepped up or down through a transformer by magnetic induction.

While you may be most familiar with alternating current from the handy outlets in your home, the ac waveform is also found as radio and video signals. Electronic circuits generally are combinations of both direct current and alternating current.

Transistors in audio amplifiers need a dc voltage to conduct current; consequently, the output of an amplifier circuit has a direct current with a superimposed ac signal.

The principles used to analyze dc circuits also apply to ac circuits. Ohm's law is still in effect. The big difference is that when the alternating voltage reverses in polarity, it produces a current that reverses in direction. Ac circuits are made up of voltages and currents constantly changing in amplitude and direction. The 120-volt outlets in homes will have an ac voltage with a positive and negative waveform (Fig. 8-1).

There are three kinds of electric power plants: fossil fuel which supplies about 78 percent of America's electricity; hydroelectric, which supplies about 10 percent; and nuclear plants, about 12 percent. They use steam to turn a turbine which turns a generator rotor producing the electricity, but the principle is still just spinning magnets inside a coil of wire. This rotating action causes the polarity of the generated voltage to alternate from positive to negative and the current to reverse directions (Fig. 8-2). The value of these varying voltages

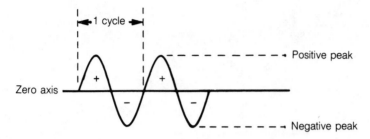

Fig. 8-1. Ac waveform with positive and negative peaks.

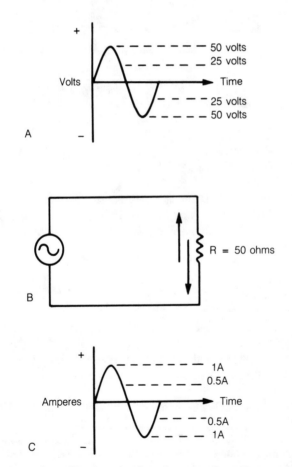

A

B

C

Fig. 8-2. An alternating voltage in a circuit produces an alternating current in the same circuit. (A) The waveform of the alternating voltage. (B) The ac circuit. (C) The waveform of the alternating current.

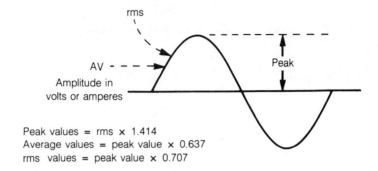

Peak values = rms × 1.414
Average values = peak value × 0.637
rms values = peak value × 0.707

Fig. 8-3. Values for a sine wave can be in voltages or current.

and currents can be classified as peak, average and root-mean-square (rms) voltages or currents (Fig. 8-3).

This alternating waveform is called a *sine wave* because to calculate the changing values, the sine (a trigonometric function of an angle) is used to determine values in circular motion. The peak value is considered the highest value of the voltage or current in a half-cycle. The average value is the mathematical average of all the values in a half-cycle. The half-cycle is used since with equal positive and negative energies the average value of a full cycle would be zero. The average value is determined by adding up all the sine values for 180 degrees (a half-cycle), then dividing by the number of sine values. This turns out to be 0.637.

The peak value of the sine is 1. It follows that the average value can be found by multiplying 0.637 times the peak value. For example, with a common peak value of 170 volts multiplied by 0.637 we get an average voltage of 108v. Root-mean-square (rms) can be thought of as the effective value of the sine wave. This turns out to be at 45 degrees or 70.7 percent of the peak value. You can now find the rms value by multiplying 0.707 times the peak value. Using the same peak value of 170V, the rms value is 0.707 times 170 volts, or 120 volts, which is present in home outlets.

TRANSFORMERS

Alternating current is used because it is the most efficient and economical method of transferring electric energy. This is possible only through the electromagnetic induction of the transformer (Fig. 8-4). The electrical energy of the supply circuit is transferred to the electrons in the secondary circuit by means of the expanding and collapsing magnetic field created by the back and forth direction of the alternating current of the supply.

Alternating current is constantly changing. It rises and falls in one direction then suddenly changes directions, rises and falls, changes direction again and so on. The magnetic field in the primary coil of the transformer is also changing in amount and direction. As this flexing magnetic field passes through the

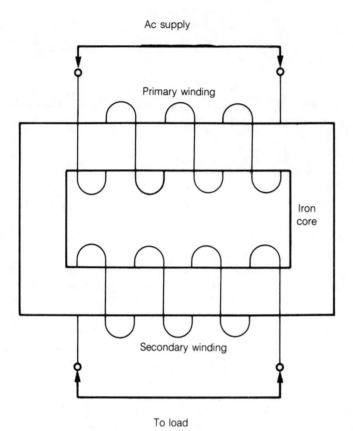

Fig. 8-4. Simple diagram of a transformer.

secondary coil, it induces a new alternating current. This new current also changes amplitude and direction at the same frequency as the supply current.

The amount of voltage developed in the secondary depends on the relative number of turns of wire in each coil and the amount of the primary voltage. For example, if the secondary coil has twice as many turns as the primary, the induced voltage in the secondary is doubled. A transformer with a 200-turn primary and a 600-turn secondary produces three times the applied voltage. In the case of a 120-volt supply, the output is 360 volts. This is called a step-up transformer. The same principle applies for a step-down transformer. Fewer turns in the secondary step the voltage down.

If you apply 120 volts to a step-up transformer and find 360 volts on the output, it would seem that 240 volts came from nowhere. A transformer cannot create energy. You cannot get more energy out of a transformer than you put in. For example, if the 360 volts on the secondary is connected to a 200 ohm load, the current in that circuit is 360 volts divided by 200 ohms (360V/200 ohms) or 1.8 amperes.

This means that the secondary circuit is using energy, or watts of power, at a rate of 1.8 amperes times 360 volts (1.8 amps × 360 volts) or 648 watts. In order for the secondary to supply the circuit, 648 watts of power must be put into the primary of the transformer. To have 648 watts in the 120 volts primary we must have 648 watts divided by 120 volts (648 watts/120 volts) or 5.4 amperes of current. The primary voltage times the current must be equal to the secondary voltage times its current, or the primary watts must be equal to the secondary watts. When the voltage is stepped up, the current is stepped down, and vice versa.

The resistance in the primary coil of a transformer is very low. Practically the only resistance to the current flow is the inductance of the windings itself. This induced voltage in the primary winding always opposes the applied voltage. It so happens that with well designed transformers, this induced voltage (counter emf) is almost equal to the applied voltage when there is no load on the secondary. Consequently, without a load the current in the primary is kept to a very low value. With no load connected to the secondary, the circuit is open and no current can flow through the windings of the secondary.

Figure 8-5 is an illustration of a step-down transformer supplying power to a 220-volt single-phase ac motor. The power source is from a convenient 2200-volt distribution circuit. Without the motor connected to the secondary,

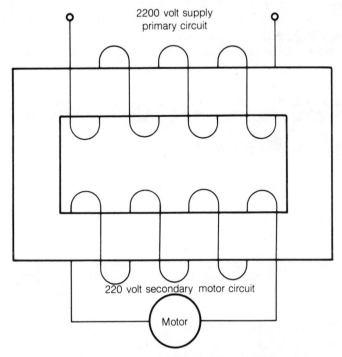

Fig. 8-5. A step-down transformer connected to a load.

the density of the magnetic field in the iron core of the transformer is just enough to induce 220 volts in the secondary windings. This is a ten to one (10:1) ratio transformer. The magnetic field was developed by the alternating current of the 2200-volt supply. With the motor connected to the secondary, the circuit is complete and current begins to flow.

If this motor requires a current of 10 amperes, this amount of current flows through each turn of the secondary winding creating a magnetic field of its own. This new magnetic field opposes the magnetic field developed by the primary winding. This causes the primary field to reduce; consequently, the counter emf is reduced, and because the 2200-volt supply is steady, more current is allowed to flow through the primary winding of the transformer. This increased current develops a magnetic force which tries to reestablish the state of the initial magnetic field. Always, the decrease in counter emf turns out to be just enough to allow the flow of the additional primary current necessary to support the load on the secondary of the transformer.

The amount of magnetic force can be thought of as proportional to the number of ampere-turns. In this example, 1 ampere of current flowing in the primary winding develops a magnetic field strong enough to counteract the demagnetizing force caused by the 10-ampere current flowing in the secondary winding. This happens because the 1-ampere primary current flows through 10 times as many turns as the 10-ampere current in the secondary of the transformer.

If you replaced the 10-ampere motor with one that required 20 amperes, the induced voltage in the secondary would instantly go to a value that would allow the primary current to increase to 2 amperes, and since there is such a large supply voltage, the current in the secondary can easily go to 20 amperes.

Generally, transformers are constant voltage devices. They supply energy at very near a constant voltage, within the limits of their rated capacity, regardless of the size of the load. Residential and commercial circuits and almost all industrial circuits are constant voltage circuits. The output voltage is maintained at a constant level and the devices—motors, lights etc.—are connected in a parallel manner.

CONSTANT CURRENT TRANSFORMERS

There are occasions where it is more practical to connect devices in series such as the circuit supplying a number of street lights.

This type of circuit is called a *varying voltage* or *constant current* circuit. While the voltage in the circuit can vary, the current must be maintained at a specific value. The more lamps that are operated, the higher the voltage must be to maintain the same current. When fewer lamps are burning, the voltage must be lowered.

This effect can be accomplished by an ingenious device called the *constant current transformer*. This transformer automatically adjusts the voltage to a level that maintains a constant current.

The constant current transformer has a primary and secondary winding where both coils are wound on the same magnetic circuit. Although both coils are mechanically free from each other, the primary winding is fixed while the secondary winding is free to move slightly. The secondary has a counterweight which almost balances the weight of the winding.

When the power is applied and all of the lines of the magnetic field in the primary sweep through the secondary the voltages developed in each turn of the coils is the same. However, if the secondary coil is moved away from the primary, fewer of the lines cut through the coil and the voltage in the secondary begins to drop. The primary winding can actually push the secondary winding away or draw it near. It does this with the magnetic field created by the current flowing in the primary. The larger the current in the primary, the farther the two coils will separate and the lower the secondary voltage.

For example, if a street lighting circuit contains 20 lamps and each lamp needs 6.6 amps, received from a voltage drop across each lamp of 72 volts, the secondary winding of the transformer is close enough to the primary to allow the magnetic field of the primary to develop (20 × 72) 1440 volts.

If five of these lamps happen to burn out and each was cleverly fitted with a device allowing the circuit to remain completed when the filament breaks, the voltage now needed for the circuit would be (15 × 72) 1080 volts. With fewer lamps, the current begins to increase. This higher current causes a greater push on the secondary winding, which moves away from the primary just far enough to allow an induced voltage of 1080 volts. This 1080 volts divided by 15 lamps gives each lamp a voltage drop of 72 volts and establishes a current flow of 6.6 amperes. The secondary will float, or balance itself in a position that maintains a constant current value regardless of the number of lamps in the circuit.

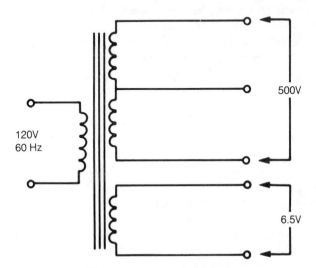

Fig. 8-6. Diagram of an iron core power transformer.

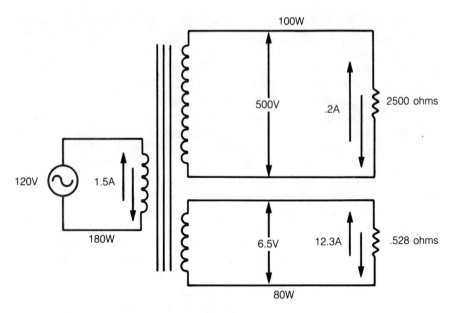

Fig. 8-7. The total power used in the two secondary circuits must equal the power in the primary circuit.

A typical power transformer has a diagram similar to the one in Fig. 8-6. There can be one, two, or more secondary windings; however, the total power in each secondary adds up to the power of the primary (Fig. 8-7).

The output of these secondary windings is an alternating current. To obtain a direct current in the output, you need to use the rectifier. Basic power supplies can use half-wave, full-wave with a center tap, and a full-wave bridge circuit (Fig. 8-8). The half-wave rectifier with just one diode only conducts during each half-cycle of the ac input (Fig. 8-9). Because it only conducts in half-cycles, the output voltage is about half of the input voltage to the circuit. This is the voltage that appears across the secondary of the transformer.

To find the average value, the voltage indicated by a dc voltmeter, you can divide the peak voltage across the secondary by 3.14 or π. For example, if there are 100 volts peak coming from the transformer, the voltage seen by the load is about 32 volts. This turns out to be one-half of the average voltage (100 V × 0.637) of the output from the transformer. A full-wave rectifier, however, allows current to flow to the load during the complete input cycle (Fig. 8-10). This circuit requires two diodes which convert both the positive and negative halves of the ac cycle into dc pulses.

By conducting on both halves of the ac signal, the output voltage of a full wave rectifier is twice that of a half-wave rectifier. In a circuit using a center tap the output voltage is always one-half of the voltage that appears across the secondary of the transformer. The full-wave bridge rectifier uses four diodes.

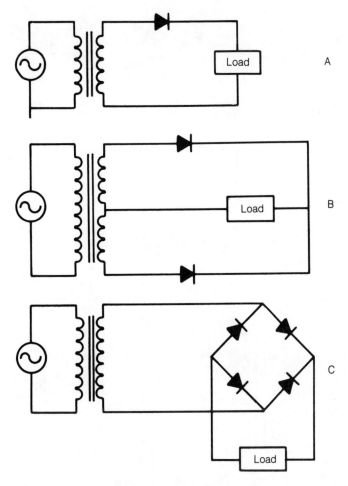

Fig. 8-8. Rectifier circuits. (A) Half-wave with input from a transformer. (B) Full-wave with a center-tap. (C) Full-wave bridge.

This clever arrangement allows the current to flow during both halves of the ac signal and delivers nearly the full voltage coming from the output of the transformer (Fig. 8-11). The output voltage of a rectifier is a pulsing dc voltage. If a standard power line is the source, the pulses of a half-wave rectifier are at the same frequency of 60 Hz (cycles) while the full-wave has a pulsating output of 120 Hz.

Most power supplies have a filter network that nearly eliminates these large voltage fluctuations (Fig. 8-12). Filter circuits vary depending on the frequency that needs to be attenuated, but generally it is the charging and discharging time of the capacitors along with some resistance value that smooths the pulsing dc voltage.

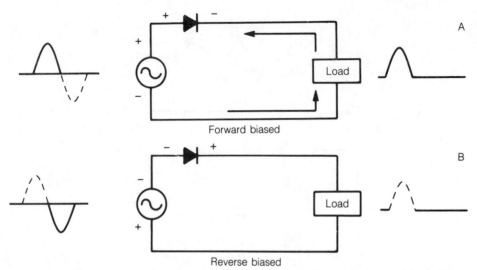

Forward biased

Reverse biased

Fig. 8-9. A half-wave rectifier conducting. (A) Circuit with current during positive half-cycle. (B) Circuit without current during negative half-cycle.

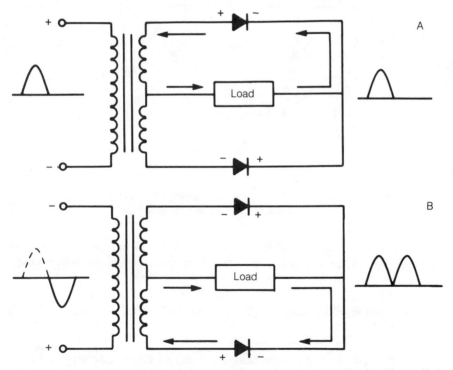

Fig. 8-10. A center-tapped full-wave rectifier conducting current. (A) Circuit with one diode conducting during positive half-cycle. (B) Circuit with second diode conducting during negative half-cycle.

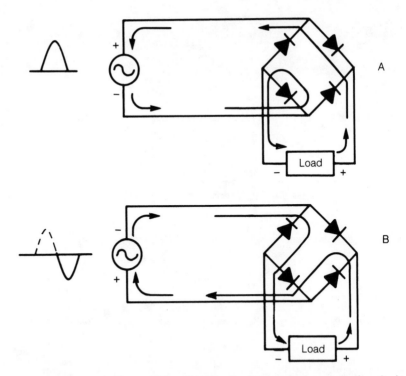

Fig. 8-11. Full-wave bridge rectifier. (A) Circuit with two diodes conducting during positive half-cycle. (B) Circuit with other two diodes conducting during negative half-cycle.

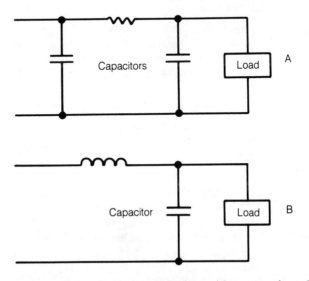

Fig. 8-12. Two forms of filter networks. (A) Resistor with two capacitors. (B) Coil with one capacitor.

Dc Motors
and Generators

Motors operate from the principle that electrical energy can be converted to mechanical energy. For generators, the principle is simply reversed.

HOW MOTORS WORK

The first law of magnetism states that like poles repel and unlike poles attract. When a current-carrying conductor is placed across a magnetic field, the magnetic field tries to push the conductor back out of the field. This is caused by the magnetic field developed from the current flow in the conductor.

If a rectangular loop of wire is placed between the poles of a permanent magnet (Fig. 9-1) nothing much happens, but if a direct current is applied to the loop a certain amount of torque is developed from the relationship between the new magnetic field and the magnet. This force can be seen from the side view in Fig. 9-2, where circles A and B represent the ends of the wire loop. This phenomena is not unlike the forces acting on an airplane wing (Fig. 9-3) where the top of the wing feels a lower pressure than the bottom. This lower pressure is what causes the wing to create lift, enabling the airplane to fly.

In Fig. 9-2 you can see that the lines of force coming from the north pole become compacted beneath the left side of the loop while they become thinned out as they continue on below the right side of the loop. Consequently, there are fewer lines above the left side and more lines concentrated above the right side. The left side is both pushed and attracted up at the same time the right side is forced down.

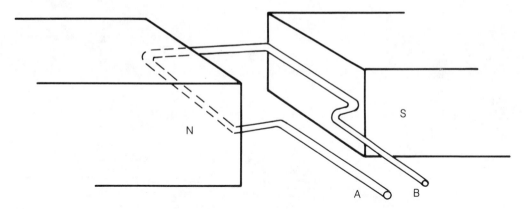

Fig. 9-1. Wire loop between magnetic poles.

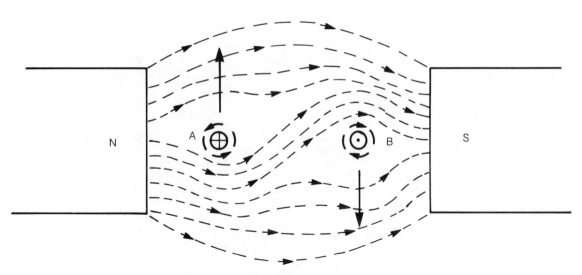

Fig. 9-2. Side view of wire loop inside magnetic field. The cross in circle A represents current flowing away while the dot in circle B represents current coming toward you (like the tail and tip of an arrow).

A closer look at the illustration explains why. The arrows circling A and B represent the direction of the magnetic field of the current in the wire, while the arrows flowing from the poles of the magnet represents that magnetic field. Above the left side of the loop, the magnetic field of the wire is in the opposite direction from the field of the magnet. This creates a thinner field. Below the left side, the lines from both fields are in the same direction. Here they combine and develop a strong magnetic field pushing the wire toward the weaker area. This combination of forces creates the torque called motor action that turns the armature in an electric motor.

111

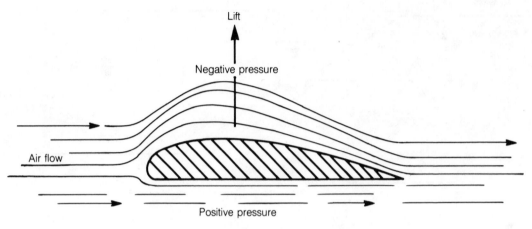

Fig. 9-3. Lift acting on an airfoil.

The wire loop in Fig. 9-2 only rotates to a vertical position and then stops because the continued pushing and lifting does not create any torque. To continue rotating, the current in the wire loop must change directions. This new direction creates a new magnetic field around the wire in the opposite direction. This

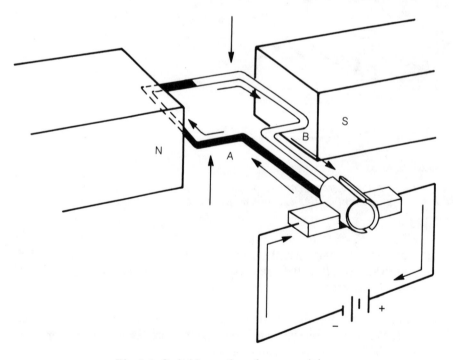

Fig. 9-4. Switching action of a commutator.

reversal can be accomplished by a simple switching arrangement called a *two-segment commutator* (Fig. 9-4).

At the position illustrated, the current flows from the negative contact through the left side of the loop and returns through the right side back to the positive connection. As the loop rotates through the vertical position, the contact that supplied the current to the left side moves from the negative connection of the commutator to the positive contact (Fig. 9-5). The right side of the loop is now touching the contact formerly used by the left side. The current and magnetic field in the loop is now flowing in the opposite direction through the loop and rotation continues. In this case, momentum must carry the loop past the vertical position. The torque in this arrangement (with a single loop) deviates greatly.

A much smoother torque can be attained by adding more loops or coils to the armature. The armature core generally consists of laminated iron (thin sheets) similar to the laminated cores of transformers (Fig. 9-6). The laminations prevent eddy currents from circulating while the motor is operating. This iron armature core makes up part of the magnetic core for the field. Slots are located around the surface of the armature providing a place to wind the coils.

The winding is simply a series of coils wound on this iron core with the ends of the coils connected to individual bars making up the commutator (Fig.

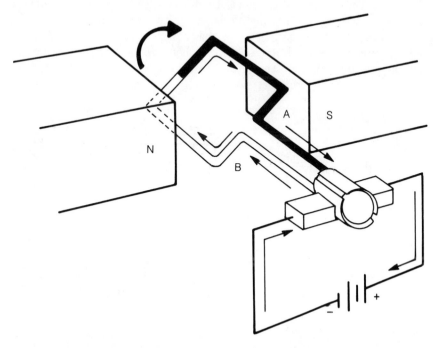

Fig. 9-5. Commutator now directing current through B half of loop to A half reversing the current flow through the conductor.

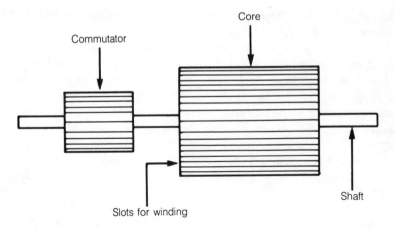

Fig. 9-6. Armature core and commutator without winding.

9-7). The number of coils and the size of the wire is determined by the horsepower, speed and operating voltage of the motor. The purpose of the armature winding is to establish motor poles on the surface of the armature core.

The copper bars of the commutator forms a cylinder around the motor shaft (Fig. 9-8). These copper bars are insulated from each other and are mounted on an insulating material attached to the shaft. The copper bars are soldered to the ends of the wires making up the coils in the armature winding. Brushes are used to complete the electrical connection between the supply, or line circuit, and the armature winding. The brushes used in electric motors are made of graphite, copper, or carbon, or some mixture of these materials.

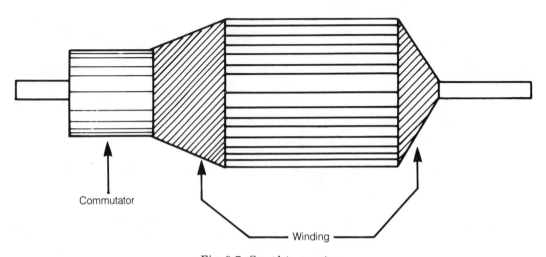

Fig. 9-7. Complete armature.

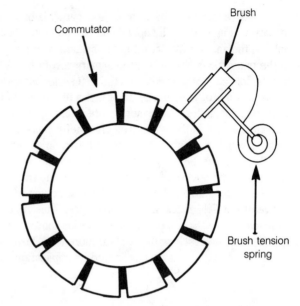

Fig. 9-8. End view of commutator.

The complete armature, the iron core, winding, commutator and shaft, is positioned inside an iron frame or housing (Fig. 9-9). Iron is used for the frame because the frame is needed to complete the magnetic circuit of the field poles. The field poles are made of iron, either solid or in laminations and support coils of wire called field windings. The field winding is an electromagnet where in this case the current for the field is supplied by the same source that supplies the armature. The iron field poles and windings complete the magnetic circuit between the frame and the armature core.

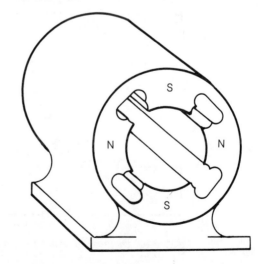

Fig. 9-9. Iron motor frame.

Field winding can be in a series, parallel, or compound circuit with the armature (Fig. 9-10). The speed and torque developed in a motor depends not only on the amount of current in the armature, it also depends on the strength of the magnetic field in which the armature rotates. A series motor is one that has its field and armature connected in a series manner (Fig. 9-10A). The field coil of this motor is constructed of just a few turns of heavy wire or may even be a strap conductor. Under normal operating conditions the field strength varies with the current in the armature. This motor has excellent starting and stalling torque, which means it will start with, or carry, very heavy overloads.

Because the field strength varies with the armature current, the torque varies with the square of the armature current. Doubling the armature current causes the field strength to be doubled, which creates four times the reaction between the field poles and the armature. This reaction produces four times as much torque. Speed control for a series motor tends to be very poor. The speed varies inversely with the load: more load, less speed and less load, more speed. Poor speed control has limited its applications, however this motor has

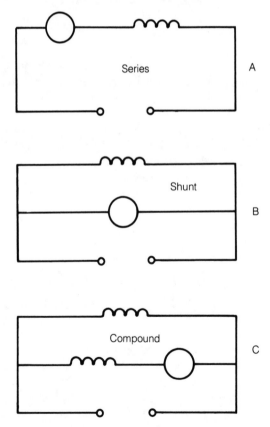

Fig. 9-10. Diagram of motor field connections.

found work as cranes and hoists were efficient load handling is accomplished with heavy loads moving slowly and light loads moving quickly.

It is important to always keep a large enough load on a series motor to keep the speed within safe limits. If the load drops to zero, the speed of the motor probably would increase to such a rate that the motor would soon destroy itself.

SHUNT MOTORS

The parallel circuit is called a shunt (Fig. 9-10B). A shunt motor tries to maintain a constant speed from no load to a full load. The field winding has many turns of small wire connected across the line or parallel with the armature winding. Because the field is connected parallel with the armature and the armature flux varies with the armature current, the field winding maintains a nearly constant magnetic field. This causes the torque of the motor to be proportional to the armature and field flux, which means the torque varies with the armature current.

For example, if a shunt motor is operating normally and the load on the motor is suddenly increased, the torque drag on the shaft is increased slowing the motor. The motor must develop more torque or stall. As the motor begins to slow, the lower rpm results in a lower emf which allows more current to flow in the armature. As the armature current increases, the torque increases to match the load. If the opposite happened and the load was suddenly dropped to zero, the increase in speed would cause an increase emf which would reduce the armature current to practically zero dropping the torque to almost zero. The speed then remains the same. The direction of rotation can be changed by reversing the leads of either the armature or the field but not both.

These motors find uses in industrial applications where a relatively steady speed is desirable with varying load conditions, such as lathes, milling machines and grinders.

COMPOUND MOTORS

The compound circuit combines the desirable features of the high torque in the series motor with the constant speed of the shunt motor (Fig. 9-10C). The compound motor has a field made up of shunt and series coils wound on each field pole. This arrangement makes the device a compound motor. Typically, the connections are made in such a manner that the series field current flows in the same direction around the field magnets as the shunt field current. The two currents flowing in the same direction produce the same polarity at each field pole and both windings aid each other in producing flux. The device now becomes a cumulative compound motor.

The torque of this motor is very good, better than the shunt motor but not as good as the series motor. The speed regulation is considered fair with the speed and torque characteristics being determined by the comparative strength of the shunt and series fields. The compound motor is suitable for industrial applications such as compressors, crushers and metal presses as well as drives for passenger and freight elevators.

Motors and generators are essentially the same machine with the same main parts. A motor converts electrical energy into rotary motion; the generator does the opposite, converting mechanical energy into electric energy. Although it will probably not be very efficient, any dc machine can be used as a motor or generator.

INDUCED EMF

Heinrich Lenz, a German-born scientist working in Russia, discovered that the induced emf in any circuit is always in such a direction that it tends to oppose the effect that produces it. Any induced current or voltage opposes the motion that causes it. If a magnet is pushed into a coil the induced current in the coil develops a magnetic field with poles such that the field repels the field of the magnet (Fig. 9-11). The same effect is felt if a magnet is abruptly removed from a coil.

Applying Lenz' law in Fig. 9-11 we know the left end of the coil must be a north pole to oppose the north end of the magnet. The direction of the induced current in the coil can be found by the left-hand rule for electron flow. If the thumb is pointed to the left toward the north pole of the coil, the fingers coil under and then over the winding in the direction of the current flow. If the magnet

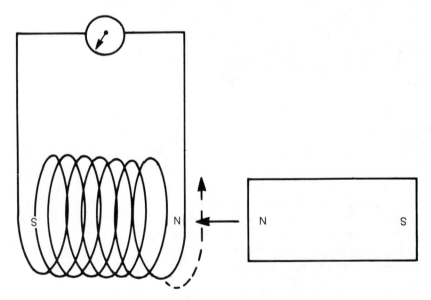

Fig. 9-11. Induced current is in direction indicated by dotted arrow. This develops a north pole at the left end of coil that opposes north end of magnet.

is then moved away, the coil must develop a south pole to oppose the movement of the magnet. Pointing the thumb of the left hand at the new north pole, which is now at the right end of the coil, again allows the curved fingers to indicate the direction of electron flow. The current has reversed and is now flowing in the opposite direction.

Michael Faraday discovered that the more magnetic lines of force, or flux, that cut across a conductor, the greater the amount of induced voltage. He also found that by increasing the number of turns in a coil, the induced voltage was also increased. It happens that the induced voltage is the sum of all the individual voltages generated in each turn in the coil. The induced voltage can also be increased by increasing the speed of the magnetic field across the conductor. This means a greater number of the magnetic lines of force are able to sweep through the conductor within a specific period of time.

Faraday's law of induced voltage states that the amount of voltage induced by a magnetic field sweeping through the turns of a coil is in proportion to the number of turns in the coil and how fast the lines of the magnetic field pass through the coils. The amount of the induced voltage can be calculated by multiplying the number of turns times the speed of the magnetic fields in webers per second. For example, if the magnetic flux sweeps through 100 turns at a rate of 2 Wb/s (100×2) you will have an induced voltage of 200V.

There is indeed a very real and effective force depicted by these unseen magnetic lines of force.

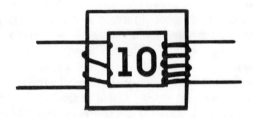

Electromagnetism in the Home

A Scottish teacher for the deaf, Alexander Graham Bell (1847-1922), arrived in Boston in 1871 and opened a school for teachers. He taught Visible Speech, a code of symbols his father had invented that indicated the position of the tongue in the throat along with the shape of the lips in producing sounds.

THE TELEPHONE

At night he conducted experiments to invent a harmonic telegraph. This device would use various electric currents to vibrate thin metal reeds which he thought would produce different sounds. He soon found that he didn't have the time or skill to build all the parts he used in his experiments. He enlisted the aid of Thomas A. Watson from a nearby electrical shop. One tedious experiment followed another. Bell thought it possible to pick up the variety of sounds of the human voice on his harmonic telegraph.

In June 1875, with Bell at one end of the line and Watson in another room at the other end of the telegraph, Bell heard the sound of a plucked reed. He immediately ran to Watson shouting, "Watson, what did you do then? Don't change anything."

They plucked reeds and listened to sounds for an hour or so then Bell gave Watson the instruction for building the first Bell telephone. He received a patent for his telephone on March 7th, 1876. Three days later Watson was the first person ever to hear words coming from a telephone. They were trying out a new transmitter in the rented top floor of a boarding house. Bell was in one room and Watson was waiting for the test message in another room when Bell spilled battery acid on his clothes. "Mr. Watson, come here. I want you!" became the first telephone transmission. Watson and Bell were good friends and eventually Watson received a share in Bell's telephone patent.

The word telephone is derived from Greek words meaning "to speak at a distance." Today the telephone is an indispensable part of most households and is the chief means of personal communications.

Bell described the principle before he invented the telephone when he said, "If I could make a current of electricity vary in intensity precisely as the air varies in density during the production of sound, I should be able to transmit speech telegraphically." This is basically what happens when you use a telephone (Fig. 10-1). Sound waves created by your voice enter the mouthpiece and strike the transmitter part of the telephone. Here the transmitter changes these sound waves to a varying electrical current. The varying current next flows over wires to a magnet in the receiver where the action of the current causes the magnet

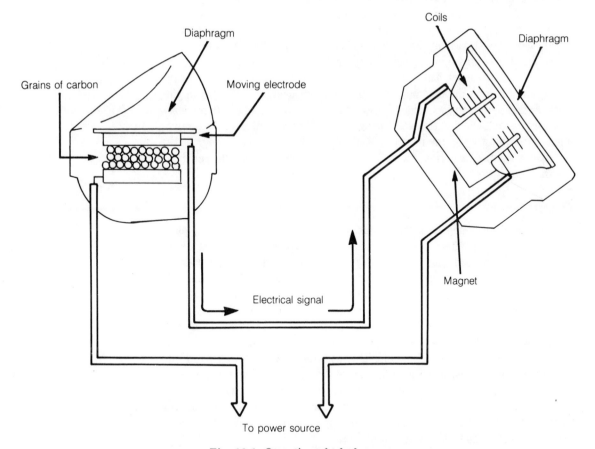

Fig. 10-1. Operation of telephone.

to act on a thin metal diaphragm. The current causes the magnet to vibrate the metal diaphragm in such a manner that the diaphragm pushes against the air creating waves duplicating the sound waves that entered the transmitter.

The mouthpiece is made up of different parts designed to capture the human voice. Your vocal cords vibrate when you speak, causing variations in air pressure, or waves similar to those caused when a pebble is dropped in a calm body of water. When these sound waves reach an ear they strike the eardrum and set it to vibrating. These vibrations are then changed into signals which are carried by nerves to the brain and we hear.

The telephone operates on the same principle. Behind the mouthpiece is the metal diaphragm. Sound waves from a voice cause the diaphragm to vibrate. A movable electrode immediately behind the diaphragm is pressing on thousands of tiny grains of carbon. The electrode applies varying degrees of pressure on the carbon granules corresponding to sound waves from the voice. Carbon is a good conductor, and with heavy vibrations the electrode presses the carbon grains tighter allowing more current to flow. Lighter vibration reduces the current. (Telephones operate from a dc power source.)

You can think of the current as vibrating in unison with the sound waves of the voice. These varying electrical impulses flow over the wires to the receiver. Inside the earpiece the fluctuating current arrives at an electromagnet. This tiny electromagnet has coils of very fine wire wound around it. As the varying current flows through the coils of the electromagnet, it develops a varying magnetic field again corresponding to the current and consequently with the voice sound waves. Attached to the electromagnet is a thin metal disc, or diaphragm, which is set vibrating by the magnetic field. This vibration pushes and pulls the air creating a duplicate series of sound waves of the ones entering the receiver, and the human voice is able ''to speak at a distance.''

TELEVISION

Along with the telephone, the television set has found a comfortable place in most homes. Magnetism plays a significant role in producing the images we see. DeForest's vacuum tube amplified radio signals in 1907, laying the foundation for today's radios. Bell provided us with the telephone. These instruments are the electronic ears that allow you to hear events taking place at some distant point. Television gives you electronic eyes to see events at some distant point.

Television works much the same way that radios do. Radios are able to operate when sound waves are changed into electromagnetic waves and then sent through the air. For television, light as well as sound is changed into electromagnetic waves. Experimenters more than a hundred years ago paved the way and by the 1920s the early theories had turned into working models, but television would not become an industry for another thirty years. Like the telephone, television comes from the Greek word *tele,* meaning far. It also uses the Latin word *videre,* which means to see.

The broadcast signal is detected or picked up by the antenna and sent to the TV set by the antenna lead-in wires. Part of the signal is processed similar to the radio and is heard through the speaker, while the video part of the signal is channeled through a high voltage section to a cathode-ray tube. This is the heart of a television set, called the *picture tube* (Fig. 10-2).

The image you see is not put on the screen as a complete picture but is presented by a series of dots scanned horizontally across the inside of the face of the tube. A vacuum is created inside the tube with an electron gun mounted inside the vacuum and in the neck of the tube. The electron gun fires a series of electrons onto the screen called the *scanning beam.* To make the beam move

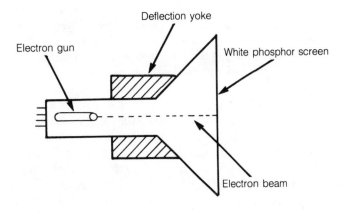

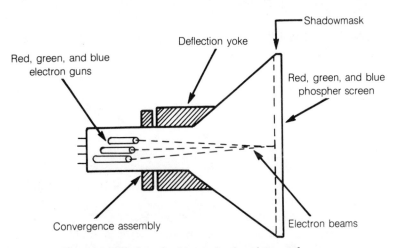

Fig. 10-2. Black and white and color picture tubes.

or scan, a *deflection yoke* is mounted on the outside of the neck of the tube. The deflection yoke is just an electromagnet that creates a magnetic field to guide the beam. Variations in this magnetic field cause the beam to sweep across the screen in horizontal lines beginning at the upper left corner of the screen, then proceeding in rapid sweeps to the lower right corner. The rapid scanning of the beam is able to produce a complete picture thirty times each second. Without the control, or guidance, of the magnetic field, the beam would assume a fixed position and quickly ruin the picture tube.

Black and white sets operate on the same principle as color. With color, however, the picture is equipped with three electron guns: one each for the colors red, blue, and green. The screen of the color picture tube is made up of hundreds of thousands of very small dots of phosphor. The electrons striking the screen cause the phosphor to light up. These phosphor dots on the screen are in clusters of three, one for each color: red, blue, and green. Electrons from the red gun strike the red dots, those from the blue gun strike the blue dots, and the green gun aims at the remaining green dots. A shadow mask behind the screen guides the three beams to their corresponding dots. Holes in the shadow mask keep the electrons from the red gun, for example, lined up with the red dots. Further dot aligning is accomplished by a convergence board. This board is mounted inside the cabinet and is fitted with coils having movable cores. These cores offer the fine tuning necessary in color alignment.

Red, blue, and green are considered primary colors and because the dots are so close together they blend together producing a true-to-life picture in full color. A few color picture tubes have been manufactured using only one electron gun. This tube doesn't use the dots but has very small red, blue, and green phosphor lines laid across the screen. An electronic switching arrangement caused the single beam to strike the individual lines in rapid succession creating the effect of a full color picture.

MAGNETIC RECORDING

The basic principle of magnetically recording sound was discovered in the late 1800s. A Danish engineer, Valdemar Poulsen, received an award for his magnetic recorder at the Paris Exposition of 1900. Early American pioneers included the experimenters at the Naval Research Laboratory and the Armour Research Institute. Magnetic recording had developed to a point that by 1930, King George V was able to record his New Year's Day greeting for broadcast by the BBC. Germany is credited with developing the first magnetic tape recording during World War II. After the war, researchers in American electronics industries developed stereophonic recording and later television recording on magnetic tape.

Wire was used as a medium for the first magnetic recording instead of tape. A thin steel wire was fed from a reel into a space between the two poles of an electromagnet. The field of the electromagnet varied with the frequency of the sound to be recorded. As the wire moved between the two poles it developed its own magnetic pattern corresponding to the frequency of the electromagnet, just like the recording head in a tape recorder.

When the wire was unwound or played back between the poles of a similar electromagnet, the magnetic pattern on the moving wire caused the electromagnet to develop a varying current with the same frequency as the magnetic pattern in the wire. This varying current from the electromagnet could be amplified and then changed back into sound impulses by a loudspeaker. The wire recorder could record with very good fidelity and could be easily erased, but it became obsolete as the more efficient and convenient magnetic tape was developed.

Today, all tape recorders, including cassettes, operate from the same basic principle (Fig. 10-3). A ribbon of magnetic tape unwinds from a feed reel at a constant speed, travels over the recording head (the electromagnet) then is wound on a take-up reel. When recording sound, a microphone changes the sound waves into a varying current. The current then travels to the recording head which in turn creates the magnetic pattern on the tape as it moves. Playback is accomplished by reversing the tape and sending the signal of the varying current to a loudspeaker instead of the microphone.

The recording can be erased by simply passing the tape across the erase head of the recorder. The erase head develops a magnetic field that realigns the magnetized molecules of the recorded pattern into their original non-magnetized form, and the tape is ready to use again. In video recording, the camera takes the place of the microphone and changes the light waves into a varying current and, just like in sound recording, the pulsating current flowing to the recording head creates the magnetic field in the head that writes the pattern on the tape.

In a TV camera, a lens focuses the light rays of a scene onto the photoelectric screen of an image orthicon tube or camera tube (Fig. 10-4). The camera tube is equipped with an electron gun. The photoelectric screen changes the light rays into electrons which flow to a plate called a target, where they form a pattern corresponding to the scene. The electron gun scans a beam of electrons across the target where the electrons striking the image pattern are bounced back in a returning beam. The returning beam of electrons is amplified to a pulsating current which is the video signal. Color cameras use three electron guns to pick up the three primary colors. Certainly video recording is more complicated than sound recording with color recording even a little more so, but the principle is the same. The sound or picture is broken down into a pulsating current which then can be fed to an electromagnet. The electromagnet creates a magnetic field that writes a pattern on a magnetic tape.

ICEMAKERS

Magnetism even helps our refrigerators make ice. A small electric motor

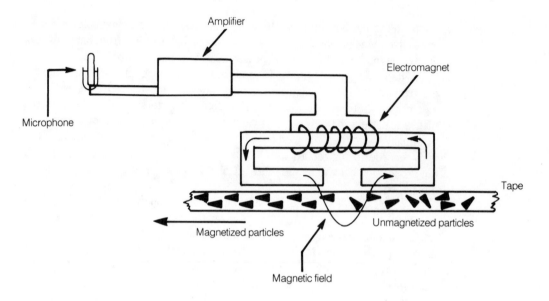

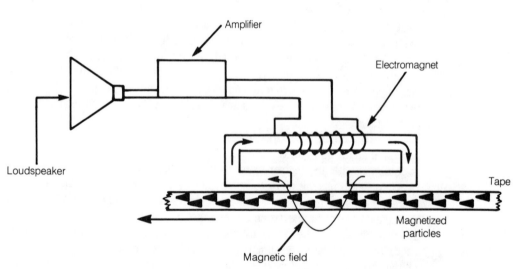

Fig. 10-3. Tape recorder.

turning a set of gears momentarily closes a switch sending a signal to a coil which develops a magnetic field around a plunger in a valve. This solenoid valve mounted in the lower part of the refrigerator, when connected to the water supply, opens briefly and allows a measured amount of water to flow into a tray in the freezing compartment. The clockworks of the system require about one hour to complete one cycle when a position on the gear activates an arm slowly emptying the tray which now is full of ice. The gear continues to turn until it once again closes the switch, beginning the cycle all over again.

TEMPERATURE CONTROL

In winter, the ducts in our homes would blow cold air when the furnace starts up if not for a small electromagnetically controlled switch in the electric circuit. Wires from the thermostat on the wall are connected to a temperature sensing device called a *thermocouple*. The wall thermostat is basically a rotating switch calibrated in numbers corresponding to the temperature. Inside the cover a small tube partially filled with mercury is positioned at an angle an is able

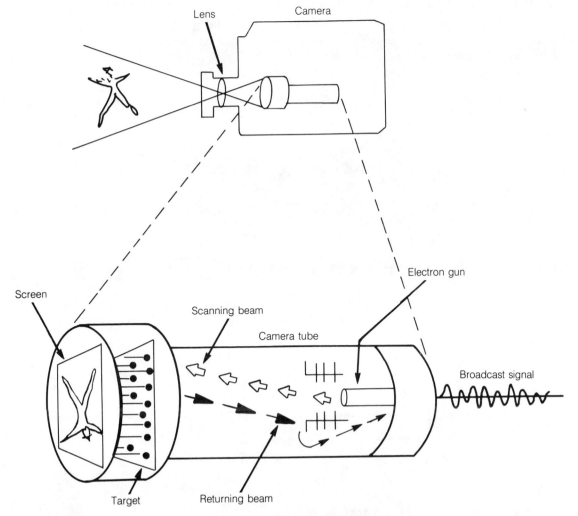

Fig. 10-4. Television camera.

to rotate slightly as the dial is moved to a temperature set point. When the room temperature drops to the set point dialed on the thermostat, the tube has tilted to such an angle that allows the mercury inside to connect two wires, like a switch, completing a circuit that pulls in a relay. This relay closes a circuit to one of the heating elements. When the element reaches a predetermined temperature another relay turns the blower on and the rest of the heaters.

When the room temperature climbs above the set point the last relay turns most the heaters off and allows the blower to continue circulating warm air. After a period of time the first relay drops out, turning off the blower along with the remaining heater.

MICROWAVE OVENS

In the kitchen the microwave oven is often taken for granted (Fig. 10-5). This electronic oven is an ingenious appliance that uses microwaves to penetrate the food and heat it. Microwaves are nothing more than radio waves generated at a high frequency. These microwaves are generated by an electron tube called

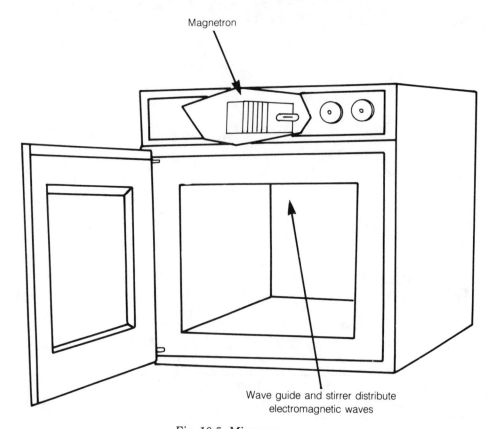

Magnetron

Wave guide and stirrer distribute electromagnetic waves

Fig. 10-5. Microwave oven.

a *magnetron*. The magnetron is an oscillator that is designed to operate in the lower range of microwave frequencies and can be found in a number of devices including radar transmitters. Basically the magnetron is a diode with a hollow cylinder for a cathode inside another hollow cylinder called an anode. Slots called resonant cavities are cut in the anode. When the cathode is heated, electrons leave its surface and travel toward the anode, but a powerful magnetic field is applied in a direction that tries to keep the electrons from the anode. As this magnetic field bends the path of the electrons, they spiral past the resonant cavities generating microwave energy. The energy then travels through a wave guide where a stirrer distributes it uniformly through the oven chamber.

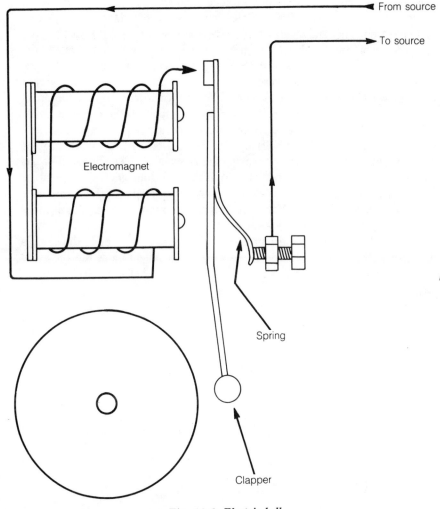

Fig. 10-6. Electric bell.

Radiation emission standards were established in late 1971 by the Department of Health, Education, and Welfare. They require that newly manufactured ovens not emit radiation over 1 millowatt per square centimeter before they are sold, or more than 5 millowatts per square centimeter throughout the working life of the oven. In addition, the door must be equipped with two safety interlocks that automatically turn off the radiation if the door is opened. The microwave oven may be one of the safest appliances in a home and is able to heat a cup of coffee in a minute and a half or a plate of leftovers in three or four.

THE DOORBELL

Magnetism is even working when someone rings your doorbell (Fig. 10-6). A pair of wires connect the push button at the door to an electric circuit that operates the bell. One wire from a power source is connected through the button to an electromagnet and an armature. (Armature here means the moving part of a relay or buzzer.) A clapper to strike the bell is mounted on the end of the armature. The armature is also fitted with a spring that presses against an electric contact made of an adjusting screw. A second wire connects the screw back to the power supply completing the circuit. When the button is pressed the circuit is completed and current flows from the power supply to the electromagnet.

Current through the electromagnet builds the magnetic field that attracts the armature, which causes the clapper to ring the bell. At the same time the spring is pulled from the screw breaking the connection. The current flow has stopped and the armature is released from the electromagnet. The armature then moves back and allows the spring to again press against the screw once more completing the circuit. If the button is still pressed, the clapper will strike the bell again and will continue to do so until the button is released. The power source may be a dry-cell battery or a bell transformer connected to the home's electrical system. Doorbells come in a variety of configurations but the same principle applies to burglar alarms, fire alarms, school bells, and factory bells.

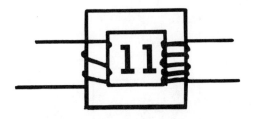

Magnetism in Science

One of the most important tools of science is the microscope. Researchers and scientists use it to study objects too small to be seen with the naked eye. The optical microscope can magnify objects because light rays are reflected from the object and are bent when they pass through one or more lenses. These bent rays create an image of the object larger than the original. Ordinary microscopes are sometimes called *light microscopes*. Although effective, these light microscopes are limited.

ELECTRON MICROSCOPES

To examine objects in much greater detail a new type of microscope was needed. X-rays were first considered, but researchers were unable to produce a lens that would control the x-rays. Scientists then found that accelerated electrons traveled with a wave motion very much like that of light and the beams could be controlled by either electrostatic or electromagnetic fields. These fields created a lens that behaved similar to the glass lens when focusing a light beam.

During the 1930s, shadow images of specimens were produced by an electron beam controlled by a magnetic lens, and by 1939 the first electron microscope came on the market. It was the *transmission electron microscope* (Fig. 11-1), which consists of a source or electron gun to provide a beam of electrons at a constant speed, a condenser lens that establishes the beam size and brightness before it strikes the specimen, a specimen stage which consists of the specimen mounted on a copper mesh grid, and an objective lens for focusing

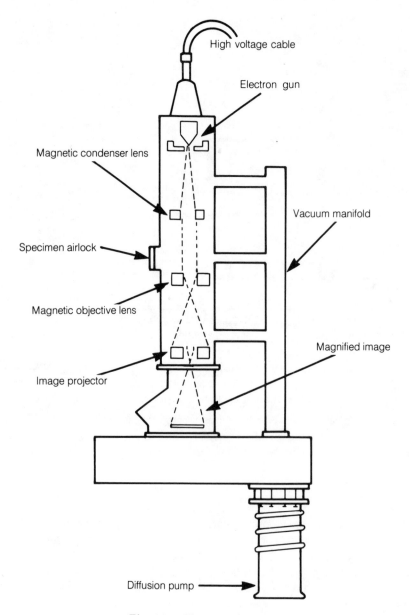

High voltage cable

Electron gun

Magnetic condenser lens

Vacuum manifold

Specimen airlock

Magnetic objective lens

Magnified image

Image projector

Diffusion pump

Fig. 11-1. Electron microscope.

and magnifying. Then the specimen is further magnified and projected on a fluorescent screen by a projector lens. The specimen can now be observed on this screen. If a permanent record of the image is needed, the screen can be replaced with a photographic plate or film.

Electrons can only travel undisturbed in a vacuum; consequently, the instrument must be evacuated to about 10^{-7} atmospheric pressure. Applying 20,000 to 100,000 volts to the electron gun achieves magnification of near 1,000,000, while useful magnification of light microscopes is about 2,000. The lens strength can be varied by varying the current. Generally electron microscopes use magnetic lenses formed by fields created by coils; however, good results have been obtained by using electrostatic lenses and magnetic lenses affected by permanent magnets.

Electron microscopes were restricted to a limited depth of field until 1965 when the scanning electron microscope became available. By 1973 both features were combined into the STEM, or scanning transmission electron microscope. This instrument provided the high resolution of the transmission system with the flexibility of the scanning electron microscope. The electron microscope has become a valuable tool used to examine a variety of metals as well as biological specimens such as viruses and cancerous tissues.

LASERS

Another remarkable piece of equipment used by scientists as well as in industry is the laser (Fig. 11-2). Lasers have been used to produce holes in very hard materials such as steel and diamonds. They are often used in welding operations.

Extremely small holes can be made without disturbing the surrounding area. Medical science has used lasers to repair detached retinas in the eyes. This condition develops when the fluid within the eyeball is able to leak through a hole in the retina. The retina can then become separated from the eyeball and

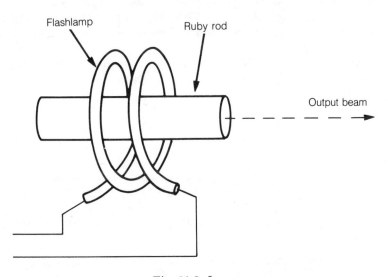

Fig. 11-2. Laser.

133

cause blindness. In the past very delicate surgery was required, but now these holes can be welded in a thousandth of a second by the beam of a laser.

Because it emits an almost perfect parallel beam, lasers find a variety of applications in measuring and surveying. Communication is another area that offers enormous potential for lasers.

Radio waves and visible light are both a form of electromagnetic radiation. The only difference is in their wavelengths. The single wavelength quality of the radiation from a laser is similar to that of a radio transmitter. Information that can be carried by electromagnetic waves is proportional to the frequency. More information can be carried as the frequency is increased. The laser beam is so much higher in frequency than radio frequencies that it might be possible for one laser beam to carry several million radio transmissions. Laser comes from the phrase "Light Amplification by the Stimulated Emission of Radiation." (Stimulated emission was first conceived by Albert Einstein in 1917.)

The atom of any substance consists of a central mass called the *nucleus*, and tiny particles called *electrons* orbit the nucleus. If the electrons orbit close to the nucleus, the atom has little energy and is in a ground state, but if the electrons circle the nucleus in a wider orbit the atom is in an excited state. The atoms are generally in a ground state, but if the material is heated, the increasing temperature raises the energy of the moving electrons which tends to move the atom to the excited state.

Once in a highly excited state, the atom tries to return to a lower energy level by giving off energy in the form of light. This light is made up of discrete packets of energy called *photons*. This is why the heated filament in a lamp produces light. This light is emitted in a wide band of wavelengths and is called *incoherent light*.

If an atom in an excited state is bombarded, for example, by a photon with the proper energy level, the atom emits another photon and lowers its energy to the proper level. The new photon is on the same wavelength as the original photon that struck the atom. There are now two photons striking atoms and they collide with the atoms in the excited stage. The process is repeated and more photons are added. Each photon produced is at the same energy level, at the same frequency, and of the same phase. This process is called *stimulated emission* and the light beam produced is monochromatic, which means that it consists of just one color rather than a mixture of several colors. Laser light is also *coherent* light because it has a single wavelength or a very narrow band of wavelengths. In normal practice, atoms in high energy states are used because atoms at lower levels tend to absorb energy and reduce the intensity of the beam.

Military science researchers have tried to develop a laser "death ray." One was developed which could bring down drone aircraft up to a couple of miles away, but these applications are limited.

PARTICLE ACCELERATION

In the early 1900s scientists discovered particles of very high energy bombarding Earth from outer space. Further study of these atomic and subatomic particles called *cosmic rays* led them to believe that the particles had been traveling in a huge magnetic field that extended throughout the galaxy. It might have taken 10 million years for these particles to build enough speed to escape this field.

The acceleration would begin when the particles collided with other moving objects in the magnetic field. In baseball, for example, a pitched ball gains speed after it strikes a bat moving toward it. If the particle strikes an object head-on it gains energy. At first thought, it might seem that if a particle gains speed in one type of collision only to lose it in another it would equal out and there would be no over-all gain, but consider a motorist on a highway. The driver sees a larger number of cars coming toward him than overtaking him. Consequently, if an equal number of objects travel in opposing directions there is more energy from head-on collisions. Given this long period of time, the particles are accelerated to a very high energy level, something in the hundreds of thousands electron volts to hundreds of millions of electron volts range.

An electron volt is a unit of work for an individual electron, not the huge number of electrons in a coulomb (which is 6.25×10.8 electrons). An electron is a charge while the volt represents a potential difference. It follows that 1 eV is the amount of work necessary to move an electron between two points that have a potential difference of one volt. Then the electron volt is simply a unit of work of charge times voltage and not current times voltage which is in watts of power.

Cosmic rays are a natural source of particles; however, they are uncontrollable and when used in research must be used as they appear in their natural state. In 1930, an American physicist, Ernest O. Lawrence, invented an atom-smashing machine called the *cyclotron*. This particle accelerator used strong magnetic fields to accelerate particles in a circular path instead of a straight line as in a linear accelerator. Basically a particle accelerator is an electrical machine that is able to accelerate charged atomic or subatomic particles to very high energy levels (Fig. 11-3). These energy levels are normally expressed in values of electron volts.

Lawrence installed two D-shaped electrodes (called dees or Ds) inside a vacuum chamber. The vacuum chamber was then placed between the poles of a powerful magnet. When a high-frequency voltage (rf) was applied to the electrodes, protons began traveling from the center of the chamber into spaces between the Ds where they were sped up by the charge created by the rapidly changing direction of the high-frequency voltage. The magnetic field contained the protons in a circular path toward the next D. The particles that crossed the space between the two Ds at the precise time the high frequency voltage was in the proper direction were boosted along their way while the particles not in phase were lost. As the particles gained speed they moved into larger orbits until they reached the limit of the magnetic field where they would break free.

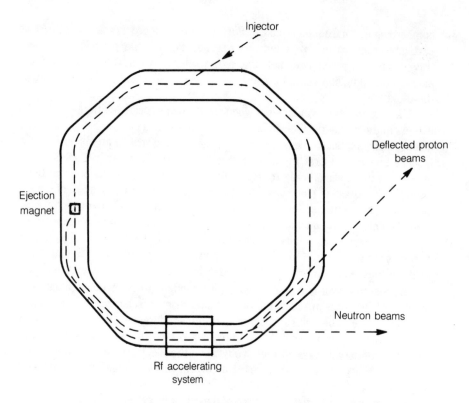

Injector

Deflected proton
beams

Ejection
magnet

Neutron beams

Rf accelerating
system

Fig. 11-3. Particle accelerator.

Lawrences' cyclotron concept is still the basis for the principal evolution of accelerator science. This led to the development of the synchrocyclotron in the 1950s, the isochronous cyclotron a decade later, and the superconducting cyclotron in the late 1970s. A couple of limitations of the earlier accelerator were discovered by H.A. Bethe and M.E. Rose in 1937. First of all, as envisioned by Albert Einstein, as any particle increases in speed it also increases in mass. The increase in mass causes the rotation rate to decrease. As the frequency of rotation decreases, the particle begins to lag until it reaches a gap where it is too late to be accelerated by the high voltage. This rf voltage will already have changed directions and starts to slow the speed of the particle. The mass increase of particles establishes limits in the process.

A second problem they found was in the magnetic field. As the particles orbit in a plane there must be some force that would push them back toward the plane if something disturbed their path. In this case this magnetic focusing was achieved by the fact that lines of the magnetic field bend in such a manner that they push down on a particle trying to rise above the plane and push up if it's drifting below. Unfortunately this restoring magnetic force is so positioned that when the accelerating particle moves to larger orbits the magnetic field is weaker and the particle begins to slow. This restricting effect is also added to the mass increase limitation.

A solution to the first problem was proposed in 1946. E.M. McMillan of the United States and V. Veksler from the Soviet Union independently thought that the answer was to adjust the frequency of the rf power to that of the slowing particle; that is, to have the frequency of the accelerating voltage in phase with the particle. This is called a synchronous phase and the idea led to the development of the synchrocyclotron.

The principle of the operation of this cyclotron is that the frequency of the voltage is synchronized with one orbiting particle. The particles that reach the accelerating gap ahead of this reference particle receive a greater voltage and energy gain. This increase in energy increases their mass and the larger mass slows their rotation, forcing them to wait for the reference particle to catch up putting the errant particles back in phase. The opposite happens to the particles reaching the accelerating gap late. They receive a lower energy increase than the reference particle and have a lower mass increase which increases their speed allowing them to catch up. The problem with this system is that lowering the frequency to that of the orbit of the reference particle causes a loss of intensity in the beam, unlike the Lawrence cyclotron where particles are accelerated on their way by exactly identical rf cycles.

To increase the beam's intensity a focusing control was required for the particles in larger orbits where the magnetic field became weaker. The basic idea was put forth as early as 1938 by L.H. Thomas but because of the complicated equations involved was not acted on for nearly 20 years. Researchers found that instead of just increasing the magnetic field, which wouldn't work, they could achieve satisfactory magnetic focusing by using wedge-shaped magnets coming together at points in the center. This ingenious arrangement of magnets was introduced in the particle accelerator called the isochronous or sector cyclotron.

Further evolution of particle accelerators came with the development of the superconducting cyclotron. Superconductivity is a term applied to a phenomenon that occurs when metals brought to extremely low temperatures (between 1 and 10 kelvin) lose all electrical resistance. This allows currents to flow through these super-cooled metals without dissipating any energy.

To get some idea of these temperatures a comfortable room temperature is about 76 degrees on the Fahrenheit scale which is about 25 degrees on the centigrade scale. Water freezes at 32 degrees F and 0 degrees C. If all of the heat energy is removed from a material, its molecules stop moving. The material is then at a temperature known as *absolute zero*. Absolute zero is a minus 273 degrees on the centigrade scale and this is the point where kelvin units start.

The basic difference in the superconducting cyclotron is that the main coil is housed in a chamber cooled by liquid helium. This superconducting coil has an increase in magnetic field strength typically three times stronger than previously possible for its size.

Particle accelerators then become much larger and more powerful with higher and higher beam energies. About 20 years ago a 400 billion electron-volt (400 GeV) accelerator, called Fermilab, was built in Batavia, Illinois. This accelerator is located underground in a circle more than a mile in diameter.

Another famous accelerator of about the same time was the efforts of several European countries for a similar facility, the CERN accelerator, near Geneva, Switzerland.

An antimatter facility at Fermilab was completed in 1985. Here scientists smash speeding antiprotons into counterrotating beams of protons. The head-on collision produces a flash of energy that gives the scientists their closest look ever at the make up of the atom. In 1984 Dr. Carlo Rubbia and Dr. Simon van der Meer of the Switzerland laboratory won the Nobel Prize for their work on an antimatter collider. Huge amounts of energy are released by collisions of matter with antimatter. In nuclear fission and fusion reactions only about 0.1 to 0.4 percent of the atomic mass is converted into energy while a reaction between matter and antimatter is 100 percent efficient. The difference between matter and antimatter is that it has the opposite electric charge. It is very rare, but experiments have proved that for every subatomic particle there exists an antiparticle. For example, for every proton an antiproton; for each electron an antielectron; and so on. Because the antiprotons are of the opposite charge they can orbit in the same magnetic focus as the protons but in the opposite direction and their collisions would produce energies near 2,000 billion electron volts (2 TeV).

Today politicians from a number of states are lobbying heavily in Washington, D.C. for a proposed 3 to 6 billion dollar physics research accelerator. The Superconducting Super Collider will be the world's largest atom smasher and have a circumference of 52 miles. It will be 10 feet in diameter and buried underground about 30 feet. This 20 trillion electron-volt accelerator will speed protons near the speed of light and will be the largest and costliest scientific device ever built. The reason it has to be so big is because as the protons increase in speed and energy they are affected by inertia. They try to travel in a straight line. The higher the energy the harder it is to make them bend, so the greater the energy of the protons, the larger the circumference of the beam. To achieve the same energies in a straight line would require a tube about 600 miles long. As machines become larger and more expensive there tends to be fewer of them. In the early 1950s the United States had over 20 accelerators; now there are only four. Project engineers have compared the project with the Egyptian Pyramids and the Great Wall of China.

The SCC will employ 10,000 electromagnets super-cooled to about 460 degrees below zero (Figs. 11-4 and 11-5). The magnets will consist of the metal alloy niobium-titanium and at that temperature they become superconductors. The cable for the magnetic coils is about ⅓ of an inch wide, consisting of 23 strands of wire. About 13 miles of superconducting wire will be used in the SSC. The magnets will guide two beams of protons in opposite directions.

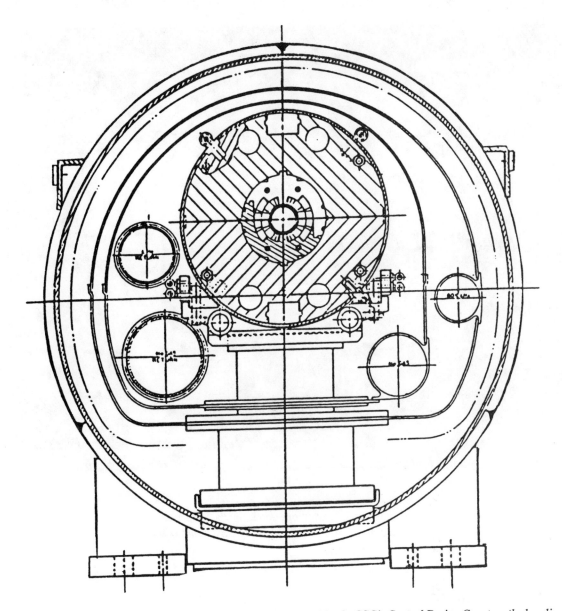

Fig. 11-4. This "D" design superconducting magnet was selected by the SSC's Central Design Group as the baseline for the SSC. A yoke magnet assembly is enclosed inside two radiation shields, layers of superinsulation, and an external vacuum vessel. (Courtesy of Westinghouse Electric Corporation.)

Within the walls of the underground tunnel, researchers would smash these two streams of high-energy protons together with a force of 40 TeV. This is the sum of energies of the two counterrotating beams. Then by analyzing the

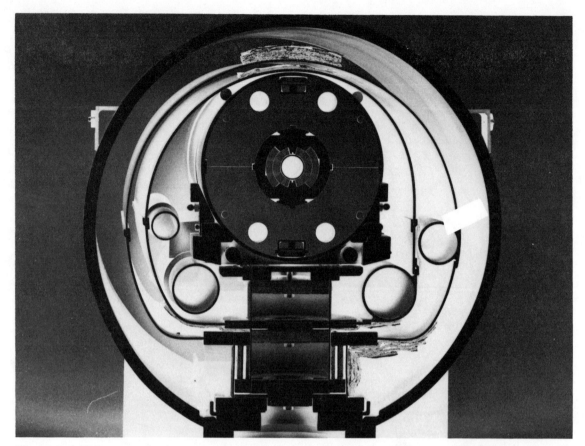

Fig. 11-5. A "D" superconducting magnet used in high energy physics research. (Courtesy of Westinghouse Electric Corporation.)

wreckage from this collision scientists would learn more about the basic character of matter and the destiny of the universe. Construction for the SSC could begin in late 1988, requiring as many as 4,500 people. It should become operational in about eight years.

SUPERCONDUCTORS

One development that could affect the construction of the SSC is the recent breakthrough in superconductors. Conventional conductors suddenly lose their electrical resistance when their temperature is lowered to absolute zero. Current flows freely without the loss of energy and very powerful magnetic fields can be developed. To achieve superconductivity, the metal, or metal alloy conductors, had to be cooled to between 1 and 10 kelvin by liquid helium. Liquid helium boils at 4.2 K (-452 degrees F) at atmosphere. By lowering the pressure, its temperature can further be reduced to 1 K. Thus, it becomes an effective cooling agent for extreme temperatures; however, helium is a rare element making this an expensive process. Liquid helium costs about $11.00 a gallon. The costs placed strict limitations on the use of superconductors. Consequently, a superconductor was needed that would conduct at a higher temperature.

By 1973, scientists had raised the temperature to 23 K using a conductor made from an alloy of niobium and germanium. Ten years later, physicist Karl Muller, in Switzerland, began to experiment with metallic oxides we call ceramics. Ceramics are normally used as insulators and are often found on high-voltage transmission lines. In 1985, Muller and associate Johannes Bednorz used a compound of barium, lanthanum, copper, and oxygen and found superconductivity would occur at 35 K. Other experimenters using similar compounds raised the temperature to 38 K. Then Paul Chu at the University of Houston, tested the compound introduced by Muller and Bednorz under pressure. The superconducting temperature raised to 52 K (-366 degrees F). After further experimenting and with the introduction of the chemical element yttrium the compound remained superconductive up to an astounding 93 K. By February, 1987, they had achieved superconductivity to 98 K. This is well within the 77 K (-320 degrees F) cooling range of liquid nitrogen, which is very inexpensive because nitrogen is a common gas, making up nearly 80 percent of the air by volume. Liquid nitrogen costs about 22 cents a gallon. Researchers are still unsure of its relationship, but oxygen seems to be the key to these rare-earth-based ceramic materials.

Normal current flow in a copper wire is limited by its conductor. Electrons constantly collide with atoms arranged in a lattice-like array in the wire and energy is lost in the form of heat (Fig. 11-6). Electrons traveling in a superconductor tend to align themselves in such a way that they pass freely through the atomic lattice work. The phenomenon occurs when a negatively-charged electron sets a slight vibration drawing the positively-charged atoms in the lattice towards it. The increased positive charge aligns the second speeding electron, allowing it to quickly follow unimpeded in the wake of the first (Fig. 11-7). In 1958 John Bardeen, Leon Cooper, and John Schrieffer presented the theory that when electrons flowed in the conventional manner, they traveled more or less freely and independently of each other. However, when flowing in a superconductor, they condense in such a way that they bond together in pairs. Bardeen, Cooper and Schrieffer were awarded the Nobel Prize in 1972 for their "BCS theory." In applying the BCS theory to the high temperature superconductors, researchers have concentrated on the lattice vibrations of oxygen atoms, as they have strong interaction with the paired electrons binding them more tightly together.

Scientists have been able to develop threadlike wires in the labs, but being ceramic, they are brittle, break easily, and are unable to carry sufficient current. With too much current, a superconductor switches back to normal conductivity.

141

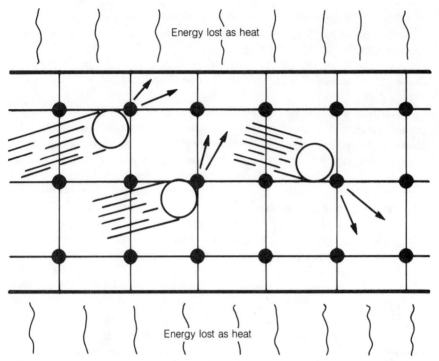

Energy lost as heat

Energy lost as heat

Fig. 11-6. Electron flow in a conventional conductor such as a copper wire. When the electrons strike the atoms in the wire energy is lost as heat and must be constantly replenished by an outside source such as a battery.

Until recently, the new materials could handle about 1200 amps per square centimeter. To be of practical use in transmission lines, magnets, and computers, current densities in the 100,000 amps per square centimeter range will be necessary. Now, scientists at IBM's Watson Research Center have developed a superconductor, with a process used in manufacturing computer chips, is able to pass a current of 100,000 amps per square centimeter. Until now superconductors were made up of grains of material packed together. Researchers felt that the boundaries between these grains could reduce their conductivity, so IBM researchers used a method called vapor deposition to develop a superconductor in the form of a pure, single crystal one inch in diameter and one micron (about ¹⁄₁₀₀th of a human hair), thick. In tests, the crystal was able to handle the 100,000 amp current density.

Scientists at the Watson Research Center have also made superconducting circuits called SQUIDS (superconducting quantum interference device). SQUIDS are able to sense minute changes in a magnetic field and are often used as high-sensitivity magnetometers. The basic elements in SQUIDS are superfast electronic switches called Josephson junctions, named after Brian Josephson, the British physicist who discovered the principle. With this capability, SQUIDS could prove invaluable in computer circuitry. IBM scientists have been able to coat suitable surfaces with the new superconductor through a method called *plasma spraying*. To create the plasma, the material (compounds of copper oxide with yttrium and barium) is heated until it is ionized. Then a thin layer is deposited on the proper surface from the plasma. With this technique, coating could be applied in straight lines like the ones used on conventional printed circuit boards as well as a variety of shapes such as rings, wires and tubes.

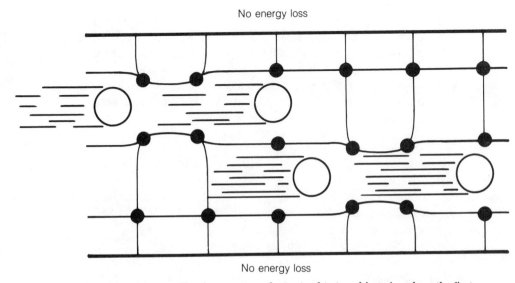

No energy loss

No energy loss

Fig. 11-7. Electrons traveling in a superconductor tend to travel in pairs where the first electron creates a wake for the second electron to be pulled through with no loss of energy.

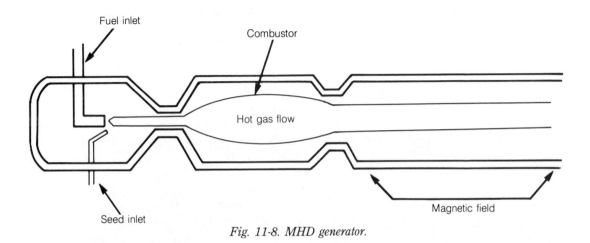

Fuel inlet

Combustor

Hot gas flow

Magnetic field

Seed inlet

Fig. 11-8. MHD generator.

A LOOK AT THE FUTURE

Although 98 degrees K (−283 degrees F) is still very cold, the rapid sequence of development and the flurry of current research leads some scientists to believe they will have compounds that can remain superconductive up to 240 degrees K (−28 degrees F) and eventually at room temperature.

If this could be achieved and the problems of brittleness and the lack of ability to handle high currents solved, superconductors could be brought out of the labs, and the possibilities would be staggering. Uses for electricity will be open for re-examination.

Electric power plants could use generators about half the size if they were equipped with superconducting magnets, or remote desert sites could contain immense solar collectors and, by employing high temperature superconductors, could transmit electricity to distant cities. If electrical current could travel without resistance, it would go on indefinitely. Current could be introduced into a superconducting ring where it would continue to flow without a power source until it was needed later. Power plants could generate electricity in the slack period, store it in huge underground superconducting rings, and bring it out to use when electrical demands are high.

Magnetic resonance imaging machines, or MRIs, are much more effective at probing the tissues of the body than x-rays; however because their superconducting magnets are cooled by helium, they are expensive to maintain. Cooling one with liquid nitrogen could save about $30,000 a year. MRIs would be inexpensive enough to be commonplace in hospitals.

Superfast trains would fly from city to city on an electromagnetic field, but magnetic levitation would not be left just to the railways. Heavy objects could be moved along assembly lines in factories. A Japanese firm, Kawasaki Heavy Industries, has experimented with a superconducting magnet on board a ship. The magnet develops a field, which in a jet-like action, forces sea water through a tube that moves the vessel along.

Electric cars could be revived. Today they can maintain highway speeds for only a couple of hours because of their high energy drain on the batteries. Superconducting electric storage rings could be mounted within the vehicle, extending its speed and range. At the same time, electric motors could produce much more horsepower for their size and weight, making electric cars competitive with present day vehicles.

The discovery of high temperature superconductors does bring problems to the plans for the SSC. It might be better to delay the construction until a smaller, more powerful, higher-temperature superconducting magnet is developed. At present, the 10,000 helium-cooled magnets operating at about 4.35 degrees K will require ten cryogenics plants costing about $129 million and the research facility itself will have an annual operating cost of about $270 million. If a new superconductor could be developed soon that could be cooled with liquid nitrogen it would reduce the physical size of the super collider and lower the initial investment as well as operating costs. However, this could be years away because strong magnetic fields are generated by high electric currents and so far, mass-produced ceramic materials are unable to handle these currents. Some researchers feel their current-carrying capacity will have to increase 100 times to be useful. Further, the ceramic superconductors are brittle and some method of winding these conductors into a coil must be developed in order to build the magnets. On the other hand, the scientists and engineers associated with the SSC project are already experienced with the low temperature superconducting magnets now in use at existing accelerators. They know how to build the SSC with the current technology, without any costly redesign or delay. Costs could soar even without accounting for inflation and the penalty for a lengthy delay could easily exceed the costs of the helium cryogenics system. With this thought in mind, it could be better to continue with present plans and begin construction.

Congress was impressed enough to allocate $40 million in 1987 for research on superconductors, and the National Science Foundation has made available $1.6 million in grants. Whether construction of the SSC is delayed or not, the development of higher temperature superconductors will be one of the most exciting scientific achievements of the 80's.

MAGNETOHYDRODYNAMIC GENERATORS

It was Michael Faraday's genius that laid the groundwork for today's electric motors and generators and it was his insight that proved that a conductor, even water, passing through a magnetic field would generate electricity. In 1831 he submerged a magnetized duct into the Thames River from the Waterloo Bridge and brought forth electricity. Nothing much was done with this amazing discovery until 1938 when the Westinghouse Company built what probably was the first magnetohydrodynamic generator.

A conventional generator produces electricity by rotating the conductor, or armature, inside a magnetic field. The MHD generator produces electricity from a conductor made of super-hot gasses speeding through a magnetic field (Fig. 11-8). Problems with heat and corrosion brought an end to the Westinghouse generator.

An experimental coal-fired MHD generator near Butte, Montana has generated enough energy to provide electric service to 15,000 homes. Here coal is used to create an extremely hot gas (about 4,500 degrees F) which is then fired through a magnetic field. Metallic particles are fed into this hot jet to increase its ability to conduct. Future MHD generators will probably use superconducting magnets to create the magnetic field. One of the advantages of the MHD generator is that there is practically no pollution.

As research equipment becomes more sophisticated and scientists probe deeper we are discovering not only more ways of harnessing magnetism's invisible force, but that it may be a significant component of life itself.

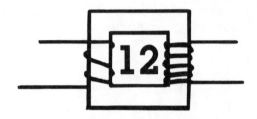

Biomagnetism

The first microscopes opened up an entirely new world full of tiny organisms and cells. The introduction of the electron microscope further exposed an even smaller world of molecules and atoms, allowing physicists to examine atomic structures. It is at this level that energy and matter become interchangeable. Scientists now suspect that magnetic fields may have a profound effect on living organisms.

EARTH'S MAGNETIC FIELD

Beginning with William Gilbert's *De Magnete* we have learned that Earth is one enormous magnet and everything on Earth exists inside an invisible magnetic force field. This dipole field is generated from the spinning liquid metal core in the center of the Earth. This unstable dynamo we live on produces the magnetic lines of force and, just like a bar magnet, these lines are more concentrated at the poles. The horizontal component is used when a compass is used as a direction finder; however, the Earth's magnetic field also has a vertical component. If a magnetic compass is turned on its side near the south pole, the needle points up. The same needle at the equator is level and at the north pole the needle points straight down.

One of the ways geologists know some of the changes the Earth has experienced is by examining the sediments laid down in the past. In the middle 1960s researchers were conducting a worldwide study of the ocean floor. Core samples containing cross sections of ancient sediments were taken to the

Lamont-Doherty Geological Observatory at Columbia University in New York City, for analysis. Here scientists determined, among other things, the strength and direction of the Earth's magnetic field at the time the sediments were laid down.

The geologists were soon amazed to find that two sections from the same core indicated opposite magnetic fields. A few thousand years had elapsed from one sample to the other but it was clear that from one period to the other Earth's magnetic field had completely reversed. It was thought at first that the reversal may have happened just in that area but, when examining core samples from around the world, they found similar patterns of a magnetic reversal. There was no doubt Earth's magnetic field had, every hundred thousand years or so, completely changed ends.

At these times the needle on a magnetic compass would have pointed south. One of the scientists, Dr. Parker, suggests that perhaps changes in speed or rotation of Earth's core could upset our dynamo enough for a reversal. First the field would be destroyed, then as the dynamo got back in step, the field would build back up but in the opposite direction, or it might just simply rotate. For now there are only theories. The fact that our magnetic field has swapped ends in the past leads researchers to believe it will do so again. One researcher, Dr. Stuart Malin, feels, based on the limited information available, that we may be at the beginning of a reversal now. It's impossible to predict, but his studies indicate a trend that if continued could cause a reversal in 250 years or so.

MAGNETISM AFFECTS LIVING THINGS

What would happen to living things if and when the next reversal occurs? Nothing much probably, but research at Lamont-Doherty may indicate otherwise. Researchers examining the sections of core samples for traces of fossilized organisms found several different species of radiolaria, a tiny form of marine life, simply disappeared about the time of the reversals. Dr. Jim Hayes' research suggests this may have not been coincidental. It seems that radiolaria often live in very deep water. Dr. Hayes concludes that if there is a connection between the organism and the magnetic field, they may use the field's vertical component as a direction finder in their migrations. A revolutionary idea to be sure, but late in 1975, Richard P. Blakemore, then a graduate student at the University of Massachusetts, discovered north-swimming bacteria.

Dr. Blakemore had been studying the salt marshes of Cape Cod when he found he could control the direction of these traveling organisms with a bar magnet. Further examination with an electron microscope revealed a row of tiny pieces of iron in each bacteria. With help from Dr. Richard Frankel of MIT he found that the iron pieces were actually tiny crystals of magnetite. It seems that each organism came supplied with its own magnetic compass.

Other forms of marine life also seem to be able to detect Earth's magnetic field. All of the species of rays, which includes the shark family, are equipped with sensitive electrical receivers. These receivers pick up an induced current when they swim through a magnetic field. At the Woods Hole Oceanographic Institute, Dr. Ad Kalmijn found he could confuse trained stingrays' ability to locate food by introducing an outside magnetic field. They could indeed sense the magnetic field. It is still not clear why they have this ability or how they use it but they are excellent navigators and certainly a compass would be handy.

Knowledge of homing pigeons was greatly improved when Dr. Charles Walcott, working with Drs. Jim Gould and Joe Kirochvink of Princeton, discovered tiny crystals of magnetite in the heads of pigeons. In their experiments they found that pigeons seem to sense the strength of Earth's magnetic field which varies with the latitudes, being the weakest at the equator. Further experiments with pigeons carrying miniature radio transmitters revealed that the birds flying into areas with a higher than normal magnetic field became disoriented and often could not escape. One such magnetic anomaly, or irregularity, in western New York, called Jersey Hill, causes pigeons to wander aimlessly, and usually never find their way home.

Jim Gould's later experiments with honeybees led him to believe that bees may use Earth's magnetic field to know what time it is. It turns out that the magnetic field varies in daily cycles and the bees may use this daily variation to set their biological clocks. Jim Gould concludes, "It could turn out that Earth's magnetic field is the second most important sense that animals in general have."

In more than 20 years of research, Dr. Frank Brown of Northwestern University has found that animals and plants alike are sensitive to magnetic fields and did respond to the ambient changes, or cycles, in Earth's magnetic field. A consultant to NASA in the 1960s, Frank Brown, did experiments which indicated that a near to zero magnetic field would result in a slight but definite change in an astronaut's nervous system. These discoveries just create more questions. If animals and plants feel a magnetic field, this sense must have existed from the very beginning of life. How are we humans affected?

MAGNETISM AND THE HUMAN BODY

In 1973 a young muskrat trapper was injured in a snowmobile accident. His right shinbone was broken in three places along with fracturing the smaller bone in the lower leg. Treatment was administered at a local hospital, but the broken bones became infected and he underwent several operations to no avail. He had tried to continue his profession by covering his full leg cast with a rubber hip boot, but in December 1976, he was admitted to the Veteran's Administration Hospital in Syracuse, New York for a possible amputation. The attending physician was Dr. Robert O. Becker who pioneered the technique of bone regeneration with extremely small amounts of electric current.

By now the young trapper's fracture had still not healed and the long cavity on the front of his shin exposed dead and infected bone. Dr. Becker's first concern was to stop the infection which included a variety of different bacteria.

First of all, bone diseases are difficult to treat because very little blood reaches the bone cells. Consequently, the body's own defense agents as well as antibiotics cannot easily get where they are needed. Secondly, no one antibiotic could kill such a variety of germs and a combination of antibiotics would likely cause more harm than good. X-rays revealed that the bones were in as bad a shape as the bacteria cultures.

Dr. Becker cleaned up the wound and after carefully removing the dead tissue and infected bone, found only a portion of the leg was left. A huge cavity existed from close to the knee nearly to the ankle. Robert Becker only used electrical treatment when it was the patient's last resort and in the past had used silver electrodes to apply the current. Coincidentally, NASA had needed an electrically conductive fabric and a small company had produced nylon parachute cloth coated with silver. Dr. Becker soaked a large piece of this silver-coated cloth in a saline solution and placed it across the wound. The piece had been cut with a strip, or tail, to hang outside the wound and serve as an electrical connection. The silver fabric was next packed in place with saline-soaked gauze. The trapper's leg was then wrapped and a dc power supply was connected. Three days later a meter on the battery pack indicated the current was starting to lower. This meant a build up of resistance at the surface of the wound.

The old dressing was removed and Becker carefully took a bacterial culture for the lab then applied a fresh silver fabric dressing. The next morning Sharon Chapin, the lab technician, showed Dr. Becker the cultures. There was a significant drop in the number of bacteria.

A few days later the young trapper was changing his own dressing and all of the bacterial cultures were sterile. Within two weeks all of the raw bone was covered with a pink coating of healing tissue and new skin was beginning to grow. X-rays taken to see the amount of bone lost revealed instead, new bone growth. Becker removed the cast and found the bones had even grown together and the happy trapper could actually lift his leg. A month later he was fitted with a walking cast and was able to leave the hospital on crutches. Two months later he returned for a check-up and walked in smiling without using crutches. The cast had suffered heavily but x-rays confirmed that the healing was almost complete.

Dr. Becker had begun his research on the ability of some animals to grow exact replacements for limbs lost in mishaps. When studying salamanders, frogs, and small mammals he discovered that electrical currents existed in areas of the nervous system. This astonishing discovery led to further understanding of the nature of bone fracture healing and limb regeneration. On an anesthetized salamander, for example, he found a pattern of varying dc potential around the body. Similar patterns are found on humans, which may offer some clue to why acupuncture can be effective. Becker suggests his method of electrical bone healing works by altering the amount of current normally flowing between these points of different potential. He further feels that the variations in Earth's magnetic field also tend to regulate these currents. There are more admissions to mental hospitals, and patients already there become even more disturbed, during magnetic storms. Even city traffic seems to worsen during blustery weather.

Dr. Becker conducted a series of experiments with magnetic fields slightly altered from our normal magnetic field. In the beginning he used men as well as animals, but when after just three generations the body weight of the exposed mice dropped to half the weight of the control, or unexposed mice, Dr. Becker stopped all further experiments on man. His experiments indicate that fields greater than, or of a different frequency component than Earth's normal magnetic field could produce signs of stress on living organisms. It seems there is a connection between the Earth's geomagnetic field and the current flowing within the nervous system of everything that is alive.

In 1983, R. Robin Baker (of the Barnard Castle experiment in Chapter 1) and his associates reported finding magnetic deposits in the spongy bone structure behind the nose of humans. Baker's research leads him to believe that even people unconsciously sense the Earth's magnetic field.

ARE MAN-MADE MAGNETIC FIELDS HARMFUL?

If what these dedicated researchers have discovered is true and all life is healthier and more natural in Earth's normal magnetic field, are there hazards for us to avoid?

Electromagnetism can be thought of as existing in two forms: fields and radiation. A field is a force in space around the object that produces it. It is a static field (that is, it doesn't vary) but if the field is varied in intensity over a time period, two fields are produced: an electric field and a magnetic field, or an electromagnetic field such as one emitted from a radio broadcasting antenna. This is electromagnetic radiation (EMR). Each radiated energy wave is made up of an electric field and a magnetic field at right angles from each other with both of these fields at right angles from the direction the wave is traveling. EMR covers a huge range of frequencies beginning with short waves of gamma rays and x-rays vibrating sextillions of times a second. This is called ionizing radiation because these rays can knock electrons from atoms, which creates very reactive ions.

At a slightly lower frequency of hundreds of trillions of times a second, the only energy visible to us is light and then the infrared waves we know as heat. As the wave gets longer we find microwaves with their billions of times per second frequency down through radio frequencies to extremely low frequencies. ELF closely relates to the dimensions of the Earth. At 10 cycles per second one ELF wave is about 18,600 miles long. Humans cannot detect any of these energies, except light and infrared heat, without some type of instrument.

For earlier civilizations there was only the weak Earth's magnetic field modulated by micropulsations from the core and further shaped by our parent star, the sun. Lightning from thunderstorms would send explosions of static beginning at about 10 kilohertz echoing over the entire world. Light alone was the largest source of electromagnetic energy. Large sections of the energy spectrum we know today were completely hushed.

The change began in 1893 with Nikola Tesla's efforts at the Chicago Worlds Fair. Then in the early 1900s Marconi, using Tesla's talents, sent his radiotelegraph message to Europe. Still our magnetic field was reasonably quiet until World War II. Shortwave radios became common and microwave radar was developed. After the war microwave telephone relay stations were established along with the microwave broadcasting of television.

It is easy to see that today we are all surrounded by a veritable sea of energies outside the normal parameters of the Earth's magnetic field. We have changed our electromagnetic surroundings more than any other area of our environment. From digital watches to flashlights and automobile ignition systems—any device that is powered by a battery produces a dc magnetic field. Over a half-million miles of high-voltage power lines are strung out over the United States producing ac or dc fields. Everyone has, at one time or another, probably noticed an annoying buzz from an AM station on their car radio near these lines. These same lines can be considered as enormous antennas broadcasting on 60 Hz in the ELF range, creating the largest radio transmitters in the world.

Low and medium frequencies are used in navigation signals and military communication along with AM radio stations. High frequencies and very high frequencies are used by CB radios, ham radios, navigation, the military, police and taxi radios, along with over ten thousand commercial radio and TV stations. The lower microwave frequencies include UHF television, some radar, weather satellites, garage-door openers, and more than ten million microwave ovens. The higher microwave frequencies are used for radar, commercial communications satellites, and about two hundred fifty thousand microwave telephone and television relay stations.

On September 14, 1983, six repairmen climbed inside an 84-foot radar tracking dish at Clear Air Force Base in Alaska. The site is part of the United States missile early warning system. During the routine maintenance a couple of turned-off flashlights began to emit light. They quickly realized that the radar dish was activated and they were in the middle of an enormous microwave oven. Two were immediately hospitalized with signs of extreme fatigue, headaches, and memory loss, among other problems. A few have since sought additional medical treatment for continuing headaches and other ailments.

In August 1985, at the Columbia-Presbyterian Medical Center in New York City, an infant girl was bombarded with magnetic and radio waves to cause the hydrogen atoms in her brain to emit their own radio signals. A computer then converted these signals to a video display of an image of the child's brain. Psychiatrists could watch a series of nine black-and-white reproductions corresponding to horizontal slices of about a fifth of an inch. Darker images indicated blood vessels and skull bone while bright white spheres represented the eyes. This new technique called magnetic-resonance imaging is used for studying the brain and researchers hope it will help solve some questions about psychiatric derangement. We can only hope that the people involved have used all their resources to study the possible side effects.

Researchers in Maryland have found that people with occupations exposing them to electricity had a much higher chance of dying of brain tumors.

Some Swedish military radar technicians experienced brain damage and an unusual protein showed up in their spinal fluid following years of exposure to high levels of microwave energy.

In Spain, researchers found that chicken embryos had undeveloped nervous systems along with hearts that didn't develop fully after being exposed only briefly to a pulsed electromagnetic field very much like those developed by video display terminals. Around the world, scientific evidence is questioning the idea that the electromagnetic fields developed by everything from our home appliances to high-voltage power lines are harmless.

Could our environment be polluted by abnormal levels of this invisible force? It may be too early to say or too late to stop, but it seems a good time to start extensive research into the hazards.

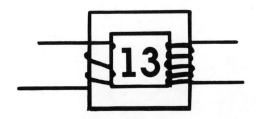

Experiments

In the third or fourth century B.C. the Chinese general Haung-ti may have been the first to use a loadstone as a compass. This first compass probably consisted of a piece of loadstone placed in a wooden bowl floating in a container of water where both the bowl and the loadstone were free to rotate.

A COMPASS

Basically the compass is a device for determining direction using a magnetic needle mounted on a pivot. The important point is that the needle is mounted in such a manner that it is free to turn and align itself with the north and south poles.

You can duplicate the general's early compass by substituting an ordinary sewing needle and a small cork for the loadstone and the wooden bowl (Fig. 13-1). A magnet must be used to magnetize the needle. Magnetize the needle by stroking it against a magnet, and then pushing it through a piece of cork in such a way that it will float in a level attitude. Next, gently place your assembly on the surface of a bowl of water. When released, the needle turns to align itself with the magnetic north and south poles. The needle is just a skinny magnet, so any nearby pieces of iron could affect the compass's direction finding ability.

A DIP INDICATOR

The vertical component of Earth's magnetic field can be observed by making a dip indicator (Fig. 13-2). To make this simple device use two needles, one

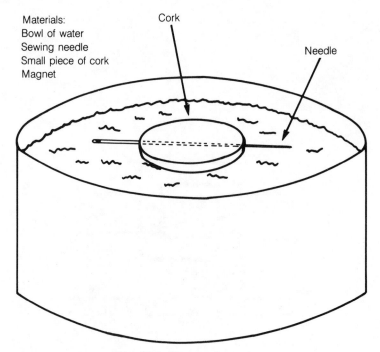

Materials:
Bowl of water
Sewing needle
Small piece of cork
Magnet

Cork

Needle

Fig. 13-1. Homemade compass.

longer than the other. Push the long needle through the center of a cork for the east-west axis. Then shove the shorter one through the center of the cork for the north-south direction. They must balance, with the short needle level, when placed on the rim of a drinking glass. Next, magnetize one end of the short needle with the south pole of a magnet. You want this end to point north, and unlike poles attract. Now place the cork and needles on the rim of the glass with the magnetized end of the short needle pointing north. If you are in an area with a strong vertical component of Earth's magnetic field, the needle dips more. If you are near Earth's magnetic equator there is no dip. At the magnetic north pole the needle points straight down.

A GALVANOMETER

In the early years an electric current was known as a *galvanic* current after Luigi Galvani. The instrument used to indicate the presence of the galvanic current was called the galvanometer. A variety of galvanometers are used in some of today's industries, but they are all based on the fact that when an electric current flows in a wire, a magnetic field is immediately established around the wire. Normally the galvanometer has a scale with numbers placed at regular intervals, but they do not represent a specific unit of current. The stronger the current, the further the needle is deflected over the scale. If the meter is

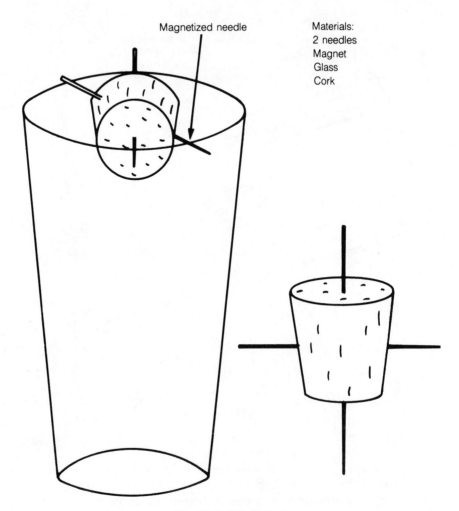

Magnetized needle

Materials:
2 needles
Magnet
Glass
Cork

Fig. 13-2. Dip Indicator.

calibrated to show standard units of current, it is known as an ammeter.

A simple device to show the presence of current can be built with just a magnetic compass and a few turns of wire (Fig. 13-3). Fold two ends of a small piece of cardboard and wrap about 30 turns of thin insulated wire around it leaving a foot or so for connections. Scrape about ½ inch of the insulation from each end of the wire. Place the compass on the cardboard and beneath the wire. Next, position the assembly so that the wires and the compass needle are at right angles to each other. The wires will be lined up in an east-west direction. When the ends of the wire are connected to a small battery the compass needle indicates the presence of an electric current. Do not leave the battery connected any longer than necessary to make observations. Variations in the needle movement

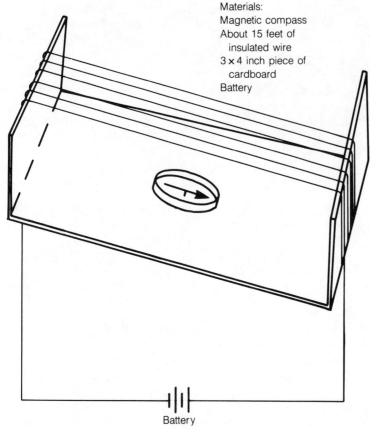

Materials:
Magnetic compass
About 15 feet of
  insulated wire
3 × 4 inch piece of
  cardboard
Battery

Battery

Fig. 13-3. Galvanometer.

can be made by reversing the connections and adding another battery.

Something to consider when adding batteries: To increase the current, batteries should be connected in parallel; that is, the negative terminal of one battery is connected to the negative terminal of the other battery with both positive terminals connected together. If the batteries are connected in series as in a flashlight, the voltage is doubled for two batteries, but the current is not increased.

INDUCING CURRENT

One of the most important features of a magnetic field is that if it travels through a coil of wire that is part of a complete circuit, it produces a current flow.

This effect can be demonstrated by moving a bar magnet through a coil connected to some device to detect the current (Fig. 13-4). Begin by constructing a coil about 2 inches in diameter. Use thin insulated wire and wrap about 20 turns for the coil leaving a length free at each end to make the connections to a meter or a home-made galvanometer. Next simply move a bar magnet abruptly in and out of the center of the coil. Notice the direction of the movement of the needle. When the magnet is first pushed into the coil the needle is deflected in one direction and then in the other direction as the magnet is withdrawn.

The greater the number of turns in the coil, the farther the needle is deflected. The alternate movement of the needle means there is an alternate direction in the flow of the current. This indicates an alternating current is present. If the magnet were able to travel in and out of the coil at the rate of 60 times a second an alternating current similar to the one in our homes would be produced.

AN ELECTROMAGNET

Permanent magnets tend to be made from hard materials, but if an intermittent magnetic field is desired, soft iron cores are placed inside a coil of wire and connected to some power source through a switch. The soft iron loses its magnetic field easily, so that this electromagnet only operates when the switch is closed. You can build an electromagnet by winding about 50 turns of thin insulated wire around a ¼ by 3 inch stove bolt (Fig. 13-5). When the ends of the coil are connected to a battery the electromagnet can pick up small nails and paper clips.

AN ELECTROMAGNETIC RELAY

The principle of the electromagnet is carried a little further to the basis

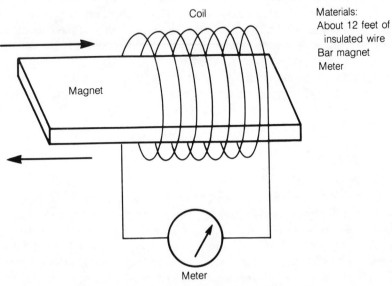

Fig. 13-4. *A magnetic field can cause a current to flow.*

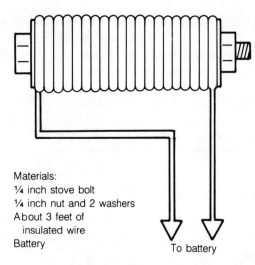

Fig. 13-5. Electromagnet.

Materials:
¼ inch stove bolt
¼ inch nut and 2 washers
About 3 feet of
   insulated wire
Battery

To battery

of a variety of electrical devices found in our daily lives, one being the electromagnetic relay. This is where electricity and magnetism working together can produce mechanical motion.

A simple relay can be used to light a lamp in a separate circuit (Fig. 13-6). First mount the electromagnet on its side to a thin board. Then wire in a switch and a 1.5 volt battery. Next push a thumb tack in each end of a wooden popsicle stick but on opposite sides. Then drive a thin nail through the edge at the center of the stick. This will be the pivot point, so some clearance must be left when mounting this wooden arm to the board. Bend two thin metal strips in the manner shown and mount to the board. These are the contacts for the lamp circuit. Wire in the lamp and second battery. Now connect the ends of the wires to the mounting screws of the contacts. Next mount the wooden arm to the board in such a way that one end is against the electromagnet and the other end presses the contacts together. When the switch is closed in the relay circuit, the wooden arm pivots, closing the contacts in the lamp circuit, and current flows to the lamp. Relays are often found when a small voltage is used to control a larger voltage in a second separate circuit. These handy devices can also be operated by a timer or a thermostat.

A SOLENOID

A solenoid is another common electromagnetic device (Fig. 13-7). Wrap several layers of paper around a large nail to make a hollow tube. Four or 5 inches of adding machine tape held by scotch tape works very well. The nail should slide freely inside the tube. Next wind about 50 turns of thin insulated wire around the tube. If the nail is positioned about halfway into the tube and the ends of the coil are connected to a small battery, the nail or plunger suddenly moves inside the tube. A larger variation of this is used to engage the flywheel of the engine when a car is started.

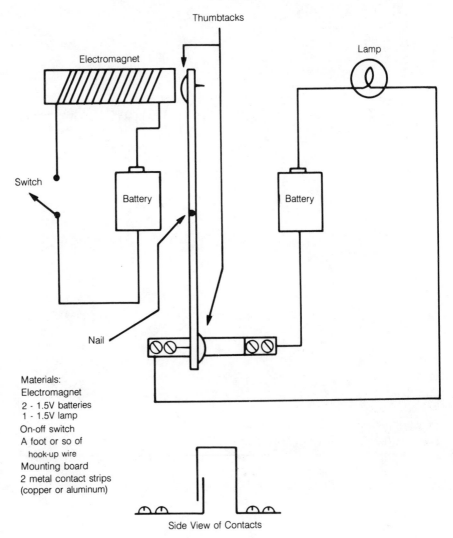

Materials:
Electromagnet
2 - 1.5V batteries
1 - 1.5V lamp
On-off switch
A foot or so of
  hook-up wire
Mounting board
2 metal contact strips
(copper or aluminum)

Fig. 13-6. Relay controlling lamp circuit.

AN ELECTRIC LOCK

Another interesting application of this principle is the electric magnetic lock (Fig. 13-8). Mount the solenoid to the fixed part of the opening and the u-shaped latch to the moving part; the lid or door. The bolt must be able to slide freely and the compression spring should only be strong enough to keep the bolt extended when no power is applied to the solenoid. Some experimenting will probably be necessary to get the lock to operate smoothly. A hidden remote pushbutton switch is used on such devices in high-security areas such as banks and government agencies.

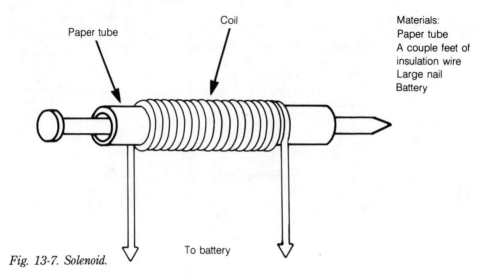

Paper tube

Coil

Materials:
Paper tube
A couple feet of
insulation wire
Large nail
Battery

To battery

Fig. 13-7. Solenoid.

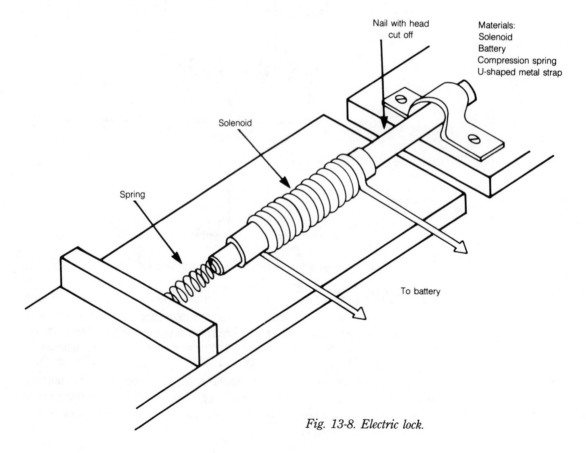

Nail with head
cut off

Materials:
Solenoid
Battery
Compression spring
U-shaped metal strap

Solenoid

Spring

To battery

Fig. 13-8. Electric lock.

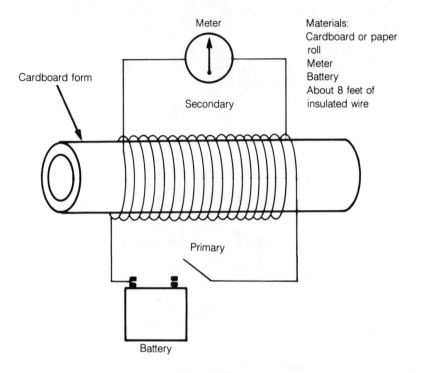

Meter

Materials:
Cardboard or paper
roll
Meter
Battery
About 8 feet of
insulated wire

Cardboard form

Secondary

Primary

Battery

Fig. 13-9. Transformer.

A TRANSFORMER

A magnetic field can also produce electrical energy. This can be demonstrated by building a simple transformer (Fig. 13-9). Wind two separate coils of wire of about 20 turns each on a cardboard form. Connect the secondary winding to an ammeter. Next, connect one lead from the primary coil to one terminal of a battery. Watch the meter and touch the remaining lead to the other terminal of the battery. The needle moves and then returns to rest, even though the battery is connected. Remove the lead and the needle again is momentarily deflected but in the opposite direction. This happens because when the lead was first touched to the battery terminal, current flowed from the battery to the primary winding. This current flow developed a magnetic field in the primary coil that, as it expanded, cut through the windings of the secondary coil. This in turn causes a current to flow in the secondary coil which is indicated by the meter movement. When the magnetic field reaches its peak and can no longer expand, the current in the secondary stops flowing. Then when the lead is removed from the battery the magnetic field starts to collapse, again cutting across the windings of the secondary coil causing a momentary flow of current, but this time in the opposite direction.

Projects

Electric motors are almost indispensable to us today and can be found performing a variety of jobs throughout our homes. Although they come in many sizes, the principle remains the same.

AN ELECTRIC MOTOR

You can see this principle in action by building a simple motor yourself (Fig. 14-1, Table 14-1). Start by cutting 8 pieces of steel or iron wire in 6½ inch lengths and wrap in a bundle (Fig. 14-2A). Next curve the ends up and trim off any uneven wires (Fig. 14-2B). This part will be the motor's field. Now wind about a half dozen layers of no. 24 enameled wire over the center of the bundle. Try to keep your winding on the taped area. Leave about a foot of each end of the coil free for connecting leads. These ends must be scraped shiny with a knife for good connections. Next press a small piece of tape through the staples for a cushion and mount the field assembly on the wood base (Fig. 14-2C).

Build the armature next by cutting 8 pieces of iron wire 2½ inches long. Form into a bundle and bind with tape near each end. Next press the nail two thirds of the way through the center of the bundle keeping an equal number of wires on each side; 4 iron wires on top and 4 on the bottom (Fig. 14-3A). After the nail, or motor shaft is centered in the bundle squeeze the wires together on each side of the shaft and bind the rest of the bundle in a layer of tape. Now the armature needs a winding.

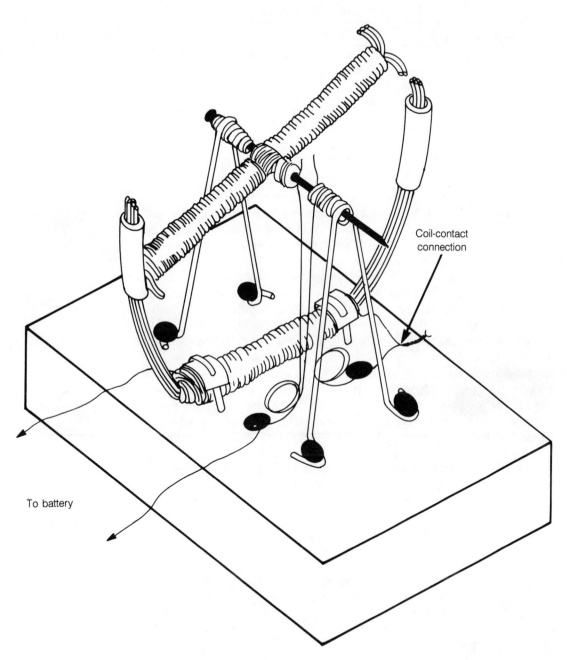

Coil-contact
connection

To battery

Fig. 14-1. Electric motor.

Table 14-1. Materials
List for the Electric Motor.

Materials:
One spool No. 24 enameled wire
7 feet of iron wire
Electrical tape
6 thumb tacks
2 staples
A nail about 2″ long
4 × 6 inch wood base

Begin next to one side of the shaft and wind (always in the same direction) outward to ¼ inch from the end, then wind back toward the shaft. Continue on over the shaft and out to ¼ inch of the other end and then back to the shaft as shown in Fig. 14-3B. Cut the coil wire leaving a few inches on each end and spread the iron wires on each end of the bundle to help keep the winding in place. Next the armature needs some kind of connection, or commutator.

Start about ¼ inch from the coil and scrape the enamel insulation from both ends of the wire for about one inch then clip off the excess wire. Then wrap the shaft with a layer of tape from the coil to about ½ inch from the pointed end and loop the scraped ends of the wires as shown in Fig. 14-3C. Position the loop contacts on each side of the shaft and secure with two narrow strips of tape (Fig. 14-3D).

Supports for the armature are made from the iron wire (Fig. 14-4A). Twist a loop in the middle to form the shaft bearing. Bend a u-shaped foot on each end and mount supports on the base with tacks. Make sure there is just enough clearance from the field to allow the armature to turn. To keep the shaft from sliding back and forth, wrap a narrow strip of tape on the head end of the shaft to form a spacer.

The contacts to the commutator are made from the enameled coil wire (Fig. 14-4B). Make sure the ends are scraped shiny and rest solidly against the commutator after mounting. Now make a connection from one commutator contact to one of the leads from the field coil. There should be two leads remaining: one from the other end of the field coil and one from the other commutator contact. These leads go to the power source. A battery or transformer will work but should not be over 6 volts or the motor will burn out. Some experimenting and adjusting will probably be required but if all the connections were well-scraped, the motor should work quite well with as little as 1½ volts from a battery.

A STEAM ENGINE

Steam engines aren't seen much anymore but the action can be recreated by building a tiny reproduction powered by a solenoid (Figs. 14-5, 14-6, Table

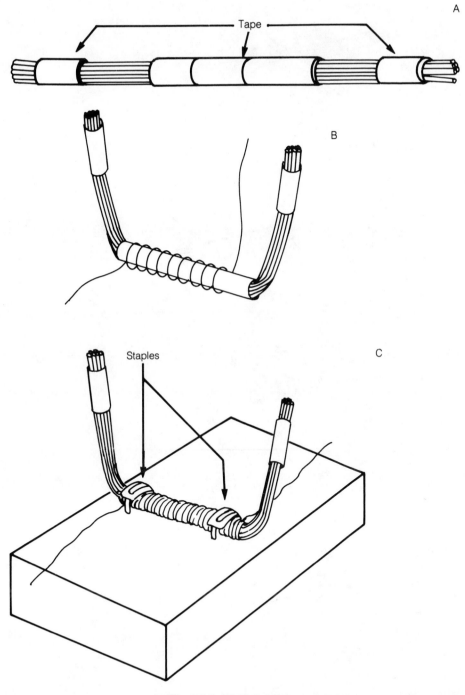

Fig. 14-2. Field for motor.

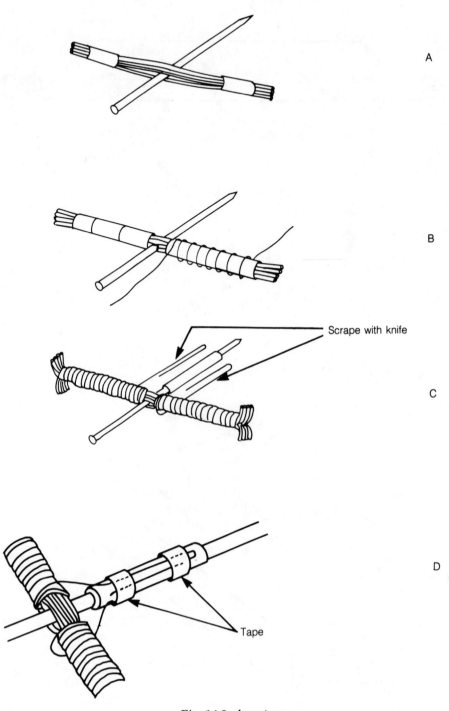

A

B

Scrape with knife

C

Tape

D

Fig. 14-3. Armature.

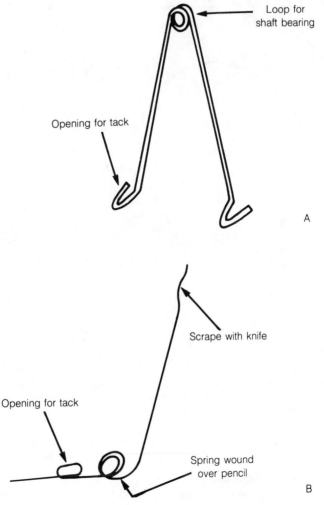

Loop for
shaft bearing

Opening for tack

A

Scrape with knife

Opening for tack

Spring wound
over pencil

B

Fig. 14-4. Armature support and contacts.

14-2). Build the 3×6 inch base and mount a solenoid in the center near one end. Try to position the center of the opening in the solenoid ¾ of an inch above the surface of the base. Heat a 10-penny nail and then let it cool to remove the brittleness. Then cut off the head and drill a ¹⁄₁₆-inch hole through the nail precisely 1⅜ inches from one end. This will be the plunger for the solenoid, or in this case, the piston rod. Mount the two 1-inch L-shaped guides on a line with the opening in the solenoid. Position one guide ¼ inch from the solenoid. Then mount the other guide 1⅜ inches from the first guide. Next mount the two 1-inch L-shaped crankshaft supports exactly 3 inches from the end of the solenoid to the center of the supports. The ¹⁄₁₆-inch bearing hole may have to be enlarged slightly to allow the crankshaft to turn freely.

167

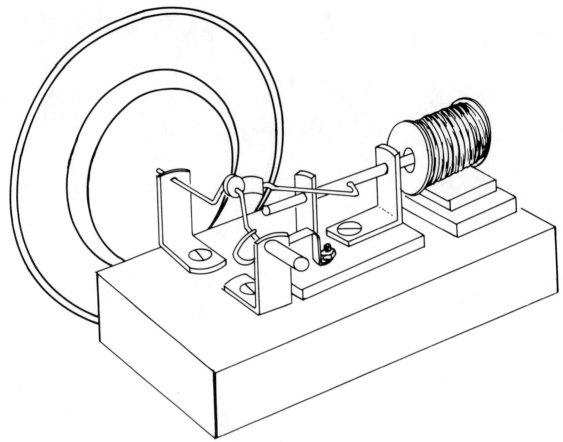

Fig. 14-5. Solenoid magnetic engine.

Table 14-2. Materials List
for the Solenoid Magnetic Engine.

Materials:
12 inches of hookup wire
2 large paper clips (for crankshaft and connecting rod)
1 solenoid
1 10-penny nail
2 ½-inch round-head wood screws
2 6-32 × 1 inch screws with nuts (for binding post)
3 small pieces of copper sheeting
4 1-inch L-shaped brackets
1 3 × 6 inch wood base
1 metal lid (for flywheel)

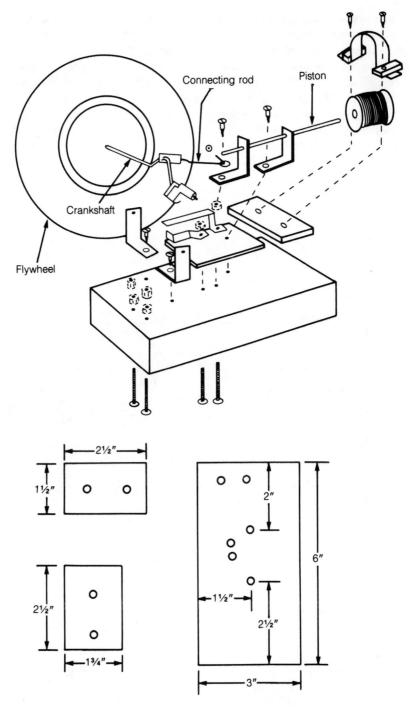

Fig. 14-6. *Exploded view of engine.*

The flywheel can be the metal bottom of an empty quart oil can or the lid from a coffee can. Punch a small hole in the exact center of the flywheel and solder one end of the crankshaft in the hole (Fig. 14-7). Now bend the connecting rod from a 2½ inch piece of ¹⁄₁₆-inch wire and solder a flat piece of brass to the crankshaft end (Fig. 14-8). Push the other end of the connecting rod through the ¹⁄₁₆ inch hole in the piston rod. This end should be held in place by two small brass or copper washers soldered to the connecting rod (Fig. 14-9). Two small washers should also be soldered to the crankshaft to form a guide for the connecting rod (Fig. 14-10).

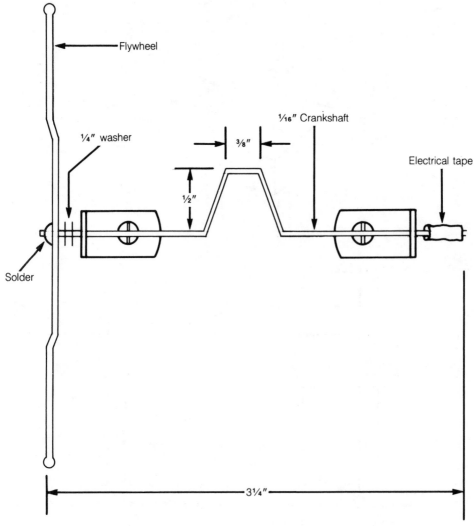

Fig. 14-7. Crankshaft.

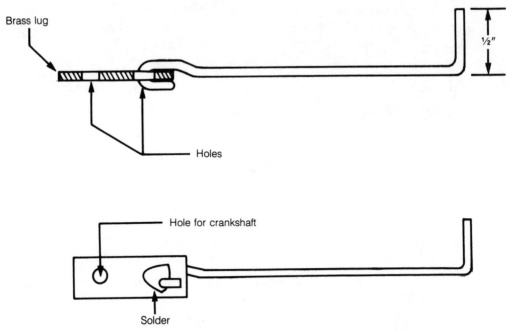

Fig. 14-8. Connecting rod.

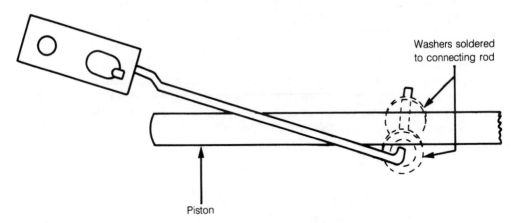

Fig. 14-9. Connecting rod to piston.

Next, solder a ⅛-inch wide 2-inch long strip of thin copper or brass sheeting to the crankshaft (Fig. 14-11). This is attached just inside the crankshaft support on the side away from the flywheel. Bend two ¼ × 2¼ inch strips of the copper sheeting to form the switch. Drill a ⅛-inch hole in each end and mount to the base with ⁶⁄₃₂ nuts as shown in Fig. 14-12.

Now mount the binding posts for the connections. Connect the wiring for

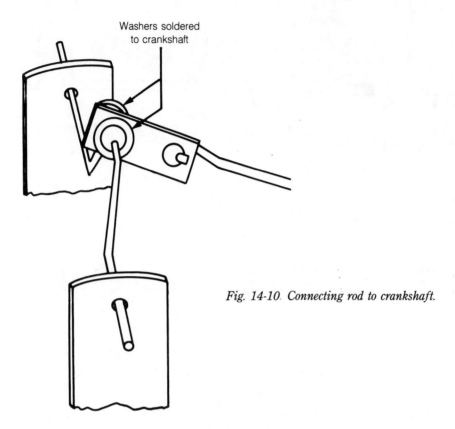

Washers soldered
to crankshaft

Fig. 14-10. Connecting rod to crankshaft.

the solenoid, switch and binding posts as illustrated in Fig. 14-13. Finally, bend the ⅛-inch strip soldered to the crankshaft so that it pushes the switch contacts together each cycle when the crankshaft is in the up position and allows the contacts to open when the crankshaft is down (Fig. 14-14). This causes the switch to open and close every half-cycle of rotation which in turn moves the piston rod in and out of the solenoid. The rotation is able to continue from the energy stored in the flywheel.

AN ELECTROMAGNETIC REPULSION COIL

Lenz's law states that an induced current will always be in such a direction that the magnetic field around the conductor will oppose the original magnetic field that induced the current. This effect can be demonstrated by a clever device called an *electromagnetic repulsion coil* (Fig. 14-15, Table 14-3).

Begin by constructing the laminated iron core. Cut several pieces of ½-inch by 6-inch strips of soft metal to form a finished iron core 9/16 inch thick. Almost any iron works except turned sheet steel and galvanized iron. Thickness isn't too important, but 1/32 inch stock works very well. Form a neat stack and secure with clamps.

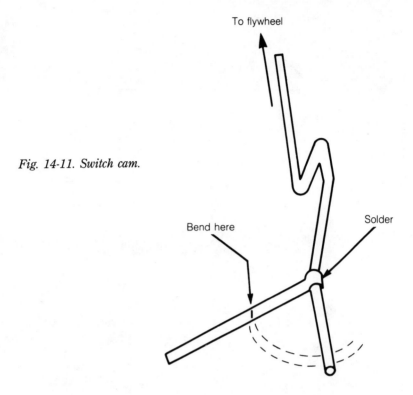

Fig. 14-11. Switch cam.

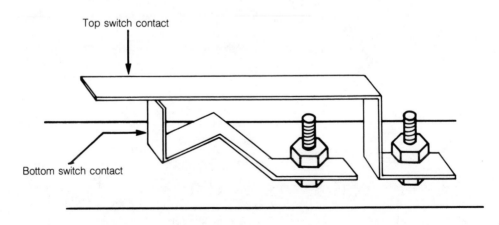

Fig. 14-12. Switch.

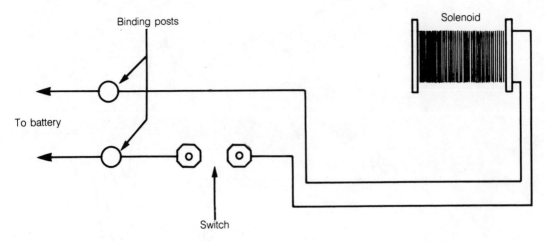

Fig. 14-13. Wiring diagram.

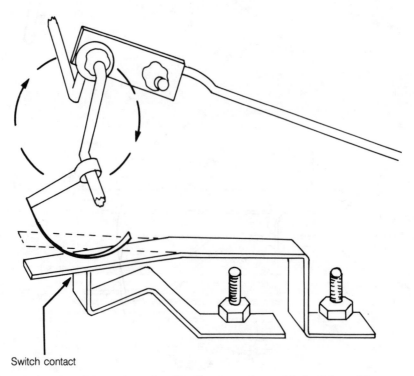

Fig. 14-14. Switch action. Contact is made when the connecting rod is in this position.

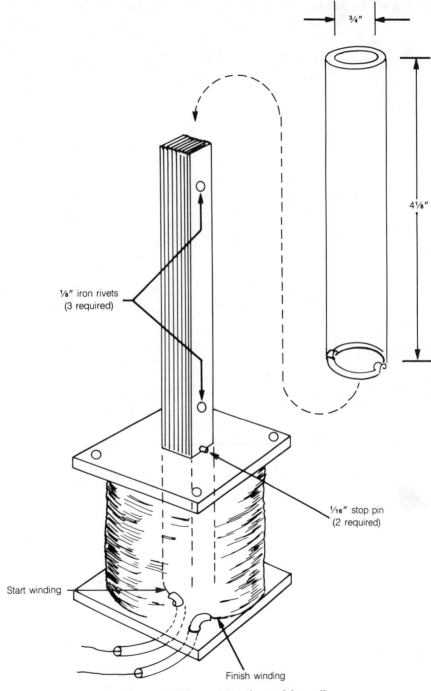

⅛" iron rivets
(3 required)

¾"

4⅛"

1/16" stop pin
(2 required)

Start winding

Finish winding

Fig. 14-15. Electromagnetic repulsion coil.

Table 14-3. Materials List for the Electromagnetic Repulsion Coil.

Materials:
Several pieces of $\frac{1}{32}$ × $\frac{1}{2}$ × 6 inch soft sheet metal
3 $\frac{1}{8}$ × $\frac{5}{8}$ inch iron rivets
2 $\frac{1}{16}$ stop pins
2 $2\frac{1}{4}$ × $2\frac{1}{4}$ × $\frac{1}{8}$ inch spool ends
1 $\frac{3}{4}$-inch inside diameter plastic tube $4\frac{1}{8}$ inches long
1 piece of armature paper or heavy wrapping paper and shellac
1 $\frac{1}{4}$ lbs. of No. 20 enameled coil wire
6 inches of spaghetti tubing
1 12-amp, 125-volt momentary contact switch
4 4-40 × $\frac{3}{8}''$ round head screws
1 wire nut
1 piece of 1 × $1\frac{1}{8}$ aluminum pipe

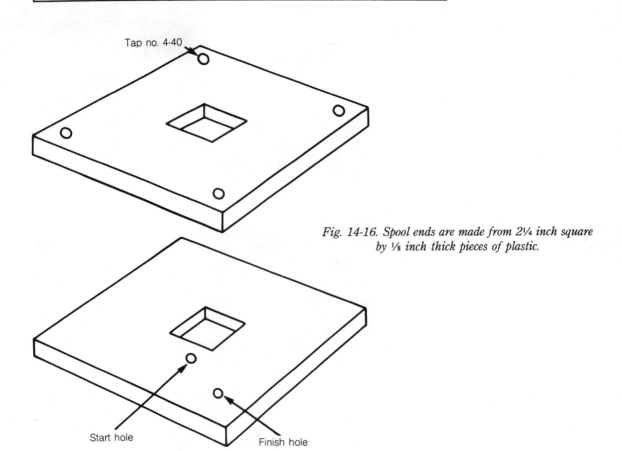

Fig. 14-16. Spool ends are made from 2¼ inch square by ⅛ inch thick pieces of plastic.

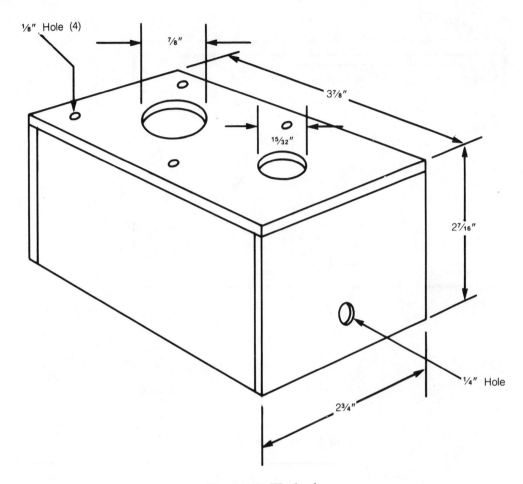

1/8" Hole (4)

7/8"

3 7/8"

15/32"

2 7/16"

1/4" Hole

2 3/4"

Fig. 14-17. Wooden base.

Next drill three 1/8-inch holes for the three iron rivets and the two holes for the pin stops. Now dress the sides and corners so that the 3/4 inch inside diameter tubing slides over the core. Cut two 2 1/4-inch squares from a piece of 1/8-inch plastic and make openings as shown in Fig. 14-16. These two pieces are the coil ends. The top one should be drilled and tapped for mounting inside the base. The bottom coil end should have the two holes for the ends of the coil winding. Both ends have the square hole for the iron core. While the top coil end is handy use it to locate the mounting holes in the base.

The base can be made from most any insulated material but 3/16 inch plywood may be the easiest to work with. Cut and assemble the pieces as shown in Fig. 14-17. Drill the 7/8-inch hole for the plastic tube and the 15/32-inch hole for the switch. Next drill the four 1/8-inch mount holes using the top coil end as a guide. A 1/4-inch hole drilled in the end by the switch can be used for the power cord.

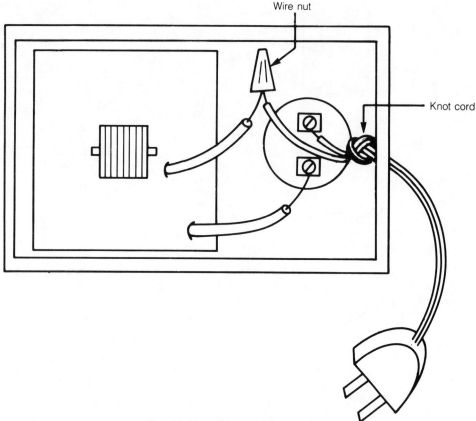

Fig. 14-18. Wiring diagram.

Next fit the two ends on the core and hold in place with the small stop pins. The space between the coil ends must now be insulated. Wrap one turn of 0.010 armature paper around the core. If this is not handy, two layers of heavy wrapping paper with a coat of shellac will work as well. The coil should be wound from #20 enameled coil wire. You should get about 40 turns to a layer and finish with 20 layers, providing a total of about 800 turns. Try to keep the turns tight and even. Cut two small notches in the plastic tube for the stop pins and attach the tube.

Mount the switch and complete the assembly. Wire the switch in series as shown in Fig. 14-18. Finally, cut the ring from a piece of 1-inch, heavy-wall aluminum pipe. This pipe will actually measure about 1⁵⁄₁₆ outside diameter and have an inside diameter of 1¹⁄₁₆ inch. This provides a wall thickness of ⅛ inch.

This coil should draw about 6 amps if there are enough turns in the winding and it is free of shorts. If desired an 8 amp fuse can easily be wired in series with the power cord. This repulsion coil will float the ring near the top of the core or shoot it into the air a couple of feet or more.

Index

Other Bestsellers From TAB

☐ **SUPERCONDUCTIVITY—THE THRESHOLD OF A NEW TECHNOLOGY**—Jonathan L. Mayo

Superconductivity is generating an excitement not seen in the scientific world for several decades! Experts are predicting advances in state-of-the-art technology that will make most existing electrical and electronic technologies obsolete! This book is the most complete and thorough introduction to this multifaceted phenomenon covering the full spectrum of superconductivity and superconductive technology. 160 pp., 58 illus.
Paper $14.95   **Book No. 3022**

☐ **LASERS—THE LIGHT FANTASTIC—2nd Edition**—Clayton L. Hallmark and Delton T. Horn

Gain insight into all the various ways lasers are used today . . . in communications, in radar, as gyroscopes, in industry, and in commerce. Plus, more emphasis is placed on laser applications for electronics hobbyists and general science enthusiasts. If you want to experiment with lasers, you will find the guidance you need here—including safety techniques, a complete glossary of technical terms, actual schematics, and information on obtaining the necessary materials. 280 pp., 129 illus.
Paper $15.95   **Book No. 2905**

☐ **FIBEROPTICS—A REVOLUTION IN COMMUNICATIONS—2nd Edition**—John A. Kuecken

Get an up-to-the-minute overview of the hottest new technology to hit the communications industry in decades! Aimed at providing a working knowledge of fiberoptic devices, this comprehensive sourcebook takes you from the basics of why and how fiberoptics were invented, right through how they work and their applications to almost any electronic purpose. You'll find dozens of practical fiberoptic applications. 352 pp., 166 illus.
Paper $22.95   **Hard $28.95**
Book No. 2786

☐ **SCIENCE FAIR: Developing a Successful and Fun Project**—Maxine Haren Iritz, Photographs by A. Frank Iritz

Here's all the step-by-step guidance parents and teachers need to help students complete prize-quality science fair projects! This book provides easy-to-follow advice on every step of science fair project preparation from choosing a topic and defining the problem to setting up and conducting the experiment, drawing conclusions, and setting up the fair display. 96 pp., 83 illus., 8 1/2″ × 11″
Paper $11.95   **Book No. 2936**

☐ **PUZZLES, PARADOXES AND BRAIN TEASERS**—Stan Gibilisco

This is a clear, concise, well-written exploration of the mysteries of the universe. It is an intriguing look at those exceptions that are as frustrating as they are amusing! The author's approach is entertaining, enlightening, and easy to understand. Although the topics are of a mathematical nature, the discussions are nontechnical. 120 pp., 83 illus.
Paper $11.95   **Book No. 2895**

☐ **BUILD YOUR OWN CUSTOMIZED TELESCOPE**—Richard F. Daley and Sally Daley

All the expert guidance you need to build an amazingly powerful telescope . . . complete with an optional computer-controlled tracking system! This guide covers site selection, building the fork mount, constructing the telescope tube, mounting the tube on the fork mount, collimating the optics, and building a rotating secondary mirror as well as instructions for adding computer-operated dual-axis tracking system using a Commodore 64 or 128 computer. 160 pp., 69 illus.
Paper $8.95   **Book No. 2656**

*Prices subject to change without notice.

Look for these and other TAB books at your local bookstore.

TAB BOOKS Inc.
Blue Ridge Summit, PA 17294-0850

Send for FREE TAB Catalog describing over 1200 current titles in print.
OR CALL TOLL-FREE TODAY: **1-800-233-1128**
IN PENNSYLVANIA AND ALASKA, CALL: **717-794-2191**